Inhalation Drug Delivery

Inhalation Drug Delivery

Techniques and Products

Paolo Colombo
Department of Pharmacy, The University of Parma, Parma, Italy

Daniela Traini
Respiratory Technology, The Woolcock Institute of Medical Research & The Discipline of Pharmacology, The University of Sydney, Sydney, Australia

Francesca Buttini
Department of Pharmacy, The University of Parma, Parma, Italy

A John Wiley & Sons, Ltd., Publication

POSTGRADUATE PHARMACY SERIES

http://www.u-l-l-a.org/

Library of Congress Cataloging-in-Publication Data

Colombo, Paolo, 1944-
    Inhalation drug delivery : techniques and products / Paolo Colombo, Daniela Traini, and Francesca Buttini.
        p. ; cm.
    Includes bibliographical references and index.
    Summary: "Provides students and those in industry with concise clear guide to the essential fundamentals in inhalation drug delivery"–Provided by publisher.
    ISBN 978-1-118-35412-4 (hardback)
    I. Traini, Daniela. II. Buttini, Francesca. III. Title.
    [DNLM: 1. Administration, Inhalation.   2. Drug Delivery Systems–methods. WB 342]
    615′.6–dc23
                                                                                    2012028075

A catalogue record for this book is available from the British Library.

Wiley also publishes its books in a variety of electronic formats. Some content that appears in print may not be available in electronic books.

Set in 10.5/13pt Times-Roman by Thomson Digital, Noida, India.

Contents

List of contributors

Francesca Buttini
Department of Pharmacy, The University of Parma, Parma, Italy

Hak-Kim Chan
Advanced Drug Delivery Group, Faculty of Pharmacy, The University of Sydney, Sydney, Australia

Gaia Colombo
Department of Pharmaceutical Sciences, The University of Ferrara, Ferrara, Italy

Paolo Colombo
Department of Pharmacy, The University of Parma, Parma, Italy

Philip Chi Lip Kwok
Department of Pharmacology and Pharmacy, LKS Faculty of Medicine, The University of Hong Kong, Hong Kong, China

David A.V. Morton
Monash Institute of Pharmaceutical Sciences, Monash University, Melbourne, Australia

Chiara Parlati
Department of Pharmacy, The University of Parma, Parma, Italy; Novartis V&D, Technology Development, Siena, Italy

Paola Russo
Department of Pharmaceutical and Biomedical Sciences, The University of Salerno, Fisciano, Italy

Rania Osama Salama
Advanced Drug Delivery Group, Faculty of Pharmacy, The University of Sydney, Sydney, Australia; Faculty of Pharmacy, Alexandria University, Egypt

Daniela Traini
Respiratory Technology, The Woolcock Institute of Medical Research & The
Discipline of Pharmacology, The University of Sydney, Sydney, Australia

Wong Tin Wui
Faculty of Pharmacy, Universiti Teknologi MARA, Puncak Alam, Selangor,
Malaysia

Paul M. Young
Respiratory Technology, The Woolcock Institute of Medical Research & The
Discipline of Pharmacology, The University of Sydney, Sydney, Australia

Series foreword

ULLA postgraduate pharmacy series

The ULLA series is an innovative series of introductory text-books for postgraduate students in the pharmaceutical sciences.

This series is produced by the ULLA Consortium (European University Consortium for Advanced Pharmaceutical Education and Research). The Consortium is a European academic collaboration in research and teaching of the pharmaceutical sciences that is constantly growing and expanding. The Consortium was founded in 1990 and consists of pharmacy department from leading universities throughout Europe including:

- Faculty of Pharmacy, Uppsala University, Sweden

- School of Pharmacy, University of London, UK

- Leiden/Amsterdam Center for Drug Research, University of Leiden, The Netherlands

- Vrije Universiteit Amsterdam, The Netherlands

- School of Pharmaceutical Sciences, University of Copenhagen Denmark

- Faculty of Pharmacy, Universities of Paris Sud, France

- Faculty of Pharmacy, University of Parma, Italy

- Faculty of Pharmaceutical Sciences, Katholieke Universiteit, Belgium

The editorial board for the ULLA series consists of several academics from these European Institutions who are all experts in their individual field of pharmaceutical science.

Previous titles include:

Pharmaceutical Toxicology

Paediatric Drug Handling

Molecular Biopharmaceutics

International Research in Healthcare

Facilitating Learning in Healthcare

Biomedical and Pharmaceutical Polymers

The titles in this ground breaking series are primarily aimed at PhD students and will also have global appeal to postgraduate students undertaking masters of diploma courses, undergraduates for specific courses, and practising pharmaceutical scientists.

Further information on the Consortium can be found at www.u-l-l-a.org

Preface

This book aims to provide a comprehensive and up-to-date understanding of the processes and mechanisms involved in inhalation drug delivery, with a strong focus on inhalation products and specific equipment and techniques used in laboratories today. It will accurately reflect the current state of our knowledge in the field of inhalation and will provide a good basis for the development of this knowledge. Theory will be covered, providing balanced new perspectives by drawing on research from a variety of fields and from industrial experience.

This book is intended as an aid to those studying pharmacy, pharmaceutical science and technology, or related subjects, at both undergraduate and postgraduate levels. Students will benefit from the concise presentation of a great deal of relevant information, and will find this book an invaluable tool for understanding the field of inhaled pharmaceutical aerosols.

1

Inhalation drug delivery

Daniela Traini

Respiratory Technology, The Woolcock Institute of Medical Research & The Discipline of Pharmacology, The University of Sydney, Sydney, Australia

1.1 Introduction

The lung offers a unique and challenging route for drug delivery for the treatment of local respiratory and systemic diseases. Advances in drug formulation and inhalation device design are creating new opportunities for inhaled drug delivery as an alternative to oral and parenteral delivery methods. Nebulizers, pressurized metered-dose inhalers (pMDIs), and dry powder inhalers (DPIs) have each found a niche in the quest for optimal treatment and convenient use. While nebulizers have evolved relatively independently of the drug formulations they deliver, the current generation of pMDIs and DPIs have been developed or tailored for the specific pharmaceutical being delivered, resulting in improved performance. However, the process of delivering drugs to the lung is not simple and is related to many factors associated with the inhaled product and the patient. This chapter will briefly review the anatomy and physiology of the lungs and the various parameters that influence drug deposition.

1.2 Brief review of the respiratory system and its physiology

The respiratory tract comprises the conducting and the respiratory regions. The conducting region essentially consists of the nasal cavity, nasopharynx, bronchi, and bronchioles. Airways distal to the bronchioles and the alveoli constitute the respiratory region, where rapid solute exchange takes place. According to Wiebel's tracheobronchial classification [1], the conducting airways comprise the first 16

Inhalation Drug Delivery: Techniques and Products, First Edition. Paolo Colombo, Daniela Traini, and Francesca Buttini.

generations, and generations 17–23 include the respiratory bronchioles, the alveolar ducts, and the alveolar sacs.

The respiratory system is made up of a gas-exchanging organ (the lungs) and a pump that ventilates it. The pump consists of: the chest wall; the respiratory muscles, which increase and decrease the size of the thoracic cavity; the areas in the brain that control the muscles; and the tracts and nerves that connect the brain to the muscles. At rest, a normal human breathes 12–15 times a minute. About 500 mL of air per breath, or 6–8 L/min, is inspired and expired. This air mixes with the gas in the alveoli, and, by simple diffusion, O_2 enters the blood in the pulmonary capillaries, while CO_2 enters the alveoli. In this manner, 250 mL of O_2 enters the body per minute and 200 mL of CO_2 is excreted.

Anatomically, the respiratory system is divided into the upper and lower respiratory tract. The upper respiratory tract consists of the nose, pharynx, and larynx. The lower respiratory tract consists of the trachea, bronchial tree, and lungs. The human respiratory tract is made up of a series of bifurcating airways, starting at the oropharynx and finishing at the alveolar sacs. The airway anatomy consists of the oro, nasopharynx, larynx, trachea, two main bronchi, five lobar bronchi (three on the right, two on the left), and 15–20 dichotomous branchings of the bronchi and bronchioles down to the level of the terminal bronchioles and the alveoli. A schematic diagram of the human respiratory tract is given in Figure 1.1.

After passing through the nasal passages and pharynx, where it is warmed and takes up water vapor, the inspired air passes down the trachea and through the bronchioles, respiratory bronchioles, and alveolar ducts to the alveoli. The portion of the airways that participates in gas exchange with the pulmonary capillary blood consists of the respiratory bronchioles and the alveoli themselves. The alveoli act as the primary gas-exchange units of the lung, especially as the gas–blood barrier

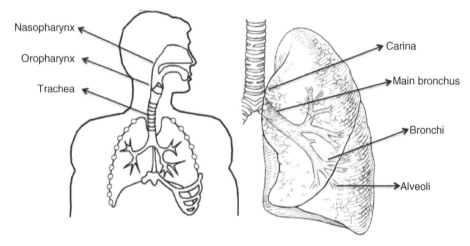

Figure 1.1 Diagrammatic representation of the structure of the human lung and airway

between the alveolar space and the pulmonary capillaries is extremely thin, allowing rapid gas exchange.

Between the trachea and the alveolar sacs, the airways divide 23 times. These multiple divisions greatly increase the total cross-sectional area of the airways, from $2.5\,cm^2$ in the trachea to $11\,800\,cm^2$ in the alveoli [2].

Consequently, the velocity of airflow in the small airways declines to very low values. Oxygen can subsequently diffuse via the alveolar epithelium (a thin interstitial space) and the capillary endothelium. In simple terms, this provides a high surface area, low surface fluid coverage, thin diffusion layer, and sluggish cell surface clearance by macrophages. These properties provide an alternative delivery system to the more conventional gastrointestinal, nasal, buccal, or transdermal delivery routes [2]. Details of the anatomy and physiology of the respiratory tract are given in many texts; the reader is referred to Moren et al. [3] or a basic anatomy text [4].

1.3 Deposition and the fate of particles in the respiratory tract

The main factors here are the properties of the aerosol particles (particle size, aerodynamic diameter) and the mode of inspiration (breath volumes, flow rate) [5]. The most important parameter defining the site of deposition of aerosol drugs within the respiratory tract is the particle characteristics of the aerosol.

Most therapeutic aerosols are almost always heterodisperse, consisting of a wide range of particle sizes and described by the log-normal distribution with the log of the particle diameters plotted against particle number, surface area, or volume (mass) on a linear or probability scale and expressed as absolute values or cumulative percentage. Since delivered dose is very important when studying medical aerosols, particle number may be misleading, as smaller particles contain less drug than larger ones. Particle size is defined from this distribution by several parameters. The mass median aerodynamic diameter (MMAD) of an aerosol refers to the particle diameter that has 50% of the aerosol mass residing above and 50% below it. Strict control of MMAD of the particles ensures reproducibility of aerosol deposition and retention within desired regions of the respiratory tract. MMAD is read from the cumulative distribution curve at the 50% point. Geometric standard deviation (GSD) is a measure of the variability of the particle diameters within the aerosol and is calculated from the ratio of the particle diameter at the 84.1% point on the cumulative distribution curve to the MMAD. For a log-normal distribution, the GSD is the same for the number, surface area, or mass distributions. A GSD of 1 indicates a monodisperse aerosol, while a GSD of >1.2 indicates a heterodisperse aerosol.

The aerodynamic diameter relates the particle to the diameter of a sphere of unit density that has the same settling velocity as the particle of interest, regardless of its shape or density. Good distribution throughout the lung requires particles with an aerodynamic diameter between 1 and 5 μm, and thus most inhaled products are formulated with a high proportion of drug in this size range [6]. In order to target the alveolar region specifically, the aerosol droplet diameter should not be more than 3 μm. Particles with diameters that are greater than 6 μm are deposited in the oropharynx, whereas smaller particles (<1 μm) are exhaled during normal tidal breathing. Particles <2.5 μm are deposited mainly in the alveoli, where they can exert no pharmacodynamic effect and are rapidly absorbed, increasing the risk of systemic adverse events. This size range was confirmed for mild asthmatics, for whom the particle size of choice should be around 2.8 μm [7].

1.4 Deposition mechanisms

As illustrated in Figure 1.2 (data from [8]), the deposition fractions of particles with different diameters in selected regions of the respiratory tract (laryngeal, upper and lower bronchial, alveolar) have been plotted as a function of particle size for particles of unit density (i.e. normalized aerodynamic diameter).

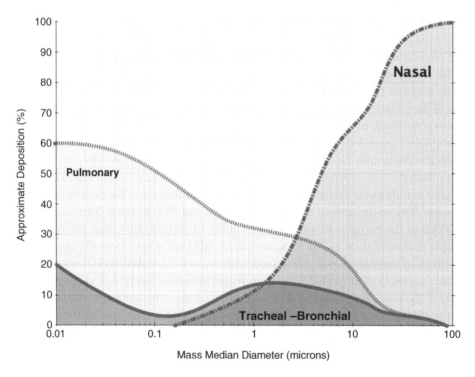

Figure 1.2 Deposition efficiency in the respiratory system as a function of the particle size

In this context, the term "deposition" refers to the mean probability of an inspired particle being deposited in the respiratory tract by collection on airway surfaces. "Total deposition" refers to particle collection in the whole respiratory tract, and "regional deposition" to particle collection in a particular region of the respiratory tract [8].

The delivery of medicaments to the respiratory tract is not a simple process. When particles do not follow airflow streamlines, coming in contact with surfaces, deposition occurs.

The entrainment and efficient delivery of particulates to the lung is regulated by three intrinsic physiological factors: inertial impaction, gravitational sedimentation, and diffusion (via Brownian motion) [9] (see Table 1.1).

Table 1.1 Deposition mechanisms in the lung and the Weibel model [10,11]. Created using data with permission from J.S. Patton and P.R. Byron. Inhaling medicines: delivering drugs to the body through the lungs. Nature Reviews Drug Discovery. 6:67–74 (2007). 11. S.W. Clarke. Medical aerosol inhalers: past present and future. In S.W. Clarke and D. Pavia (eds.), Aerosola and the lung: Clinical and experimental aspects, Butterworths, London, 1984, pp. 1–18.

Generation	Zone	Name	Diameter (cm)	Total cross-sectional area (cm^2)	Mechanism of deposition
0		Trachea	1.8	2.54	Impaction (inertia)
1		Main bronchi	1.22	2.33	
2			0.83	2.13	
3			0.56	2.00	
4			0.45	2.48	
5		Bronchi	0.35	3.11	Sedimentation (gravity)
6			0.28	3.96	
7	Conducting/ tracheobronchial		0.23	5.10	
8			0.186	6.95	
9			0.154	9.56	
10			0.130	13.4	
11			0.109	19.6	
12			0.095	28.8	
13		Bronchioles	0.082	44.5	
14			0.074	69.4	
15			0.066	113	
16		Terminal bronchioles	0.060	180	
17		Respiratory bronchioles	0.054	300	
18			0.050	534	
19	Alveolated/ respiratory	Alveolar ducts	0.047	944	Brownian diffusion
20			0.045	1600	
21		Alveolar	0.043	3220	
22		sacs	0.041	5880	
23			0.041	11800	

Although deposition occurs throughout the airways, inertial impaction usually occurs in the first 10 generations of the lung, where air velocity is high and airflow is turbulent [12]. Most particles >10 μm are deposited in the oropharyngeal region, with a large number impacting on the larynx, particularly when the drug is inhaled from devices requiring a high inspiratory flow rate (DPIs) or when it is dispensed from a device at a high forward velocity (MDIs) [13]. The large particles are subsequently swallowed and contribute minimally, if at all, to the therapeutic benefit.

Inertia is the inherent property of a moving mass that resists acceleration. It depends not only on the particle density and the particle diameter, but also on the airflow velocity.

Sedimentation is the gravitational settling of particles and mainly affects particles in the size range 1–8 μm. In this case, the distance that a particle will settle in a given time increases with the mass and is a gravity-dependent process. The longer a particle remains in the respiratory system, the larger the settling distance the particle covers and hence the greater the probability that the particle will get into contact with the airspace wall.

Brownian diffusion is the irregular motion of an aerosol particle in still air, caused by random variations in the incessant bombardment of gas molecules against the particles. It affects smaller particles (<1 μm), which deposit mainly in the alveolar region. In this area, air velocity is negligible, and thus the contribution to deposition by inertial impaction is nil. Particles in this region have a longer residence time and are deposited by both sedimentation and diffusion. Particles not deposited during inhalation are exhaled. Deposition due to sedimentation affects particles down to 0.5 μm in diameter, whereas below 0.5 μm, the main mechanism for deposition is diffusion.

In summary, minimal deposition occurs in the size range between 0.1 and 1 mm, because neither impaction, sedimentation, nor diffusion is effective in particle displacement. With decreasing particle diameter, diffusional particle displacement increases, so that particle deposition in the respiratory tract increases. With increasing particle diameter, the distance covered by sedimentation or impaction increases, so that the total particle deposition is also enhanced. The optimum size range of particulates for inhalation therapy has been shown to be 2.5–6 μm [2].

1.5   Parameters influencing particle deposition

When designing and formulating a delivery system, the many factors that influence the deposition of drug particles need to be considered.

Increasing air velocity increases impaction deposition but decreases sedimentation and diffusion deposition. Breathing patterns – tidal volumes, respiratory time, and flow rates – all influence deposition. Total respiratory tract deposition increases with mean respiratory time and tidal volume (maximum inspiration volume). Total deposition has been shown to be dependent on the mean residence time (Tm) and the tidal volume (Vt) according to the following equation [14]:

$$TDF = (DT_m)^{0.5} V_t^{0.49} \qquad (1.1)$$

where D is the diffusion coefficient of the particles in air. The respiratory period and mean residence time (Tm) of particles in the respiratory tract and the tidal volume (Vt) are the two most important breathing parameters affecting deposition of particles. Deposition increases to almost the same extent with an increase in Tm or Vt [14]. Age does not influence substantially the deposition patterns of particles, except perhaps for very small particles (1–2 nm) and for very young subjects (3 years old, in vitro cast data) [15]. There are inter-individual differences in the human population that affect the deposition and clearance, due to factors like age, existing respiratory conditions, state of the mucous layer, and exposure to other respiratory hazards (i.e. cigarette smoke) [16].

1.6 The clearance of deposited particles

Like other parts of our organism, the lung has evolved to prevent the invasion of unwanted airborne particles into the body. Airway geometry, humidity, and clearance mechanisms contribute to this elimination process. The challenge in developing therapeutic aerosols is to produce one that eludes the lung's various lines of defense.

1.7 Airways geometry and humidity

Progressive branching and narrowing of the airways encourages impaction of particles. The larger the particle size, the greater the velocity of incoming air, the greater the bend angle of bifurcations, and the smaller the airway radius, the greater the probability of deposition by impaction [17]. It must be remembered that in deposition studies, aerosol droplets and particles are highly dynamic systems [18]. The lung has a relative humidity of approximately 99.5%. Particle size does not remain constant once it reaches the respiratory tract. Volatile aerosols become smaller with evaporation, hygroscopic aerosols grow bigger with moisture from the respiratory tract, and particulates aerosols may agglomerate (Figure 1.3).

Therefore, knowledge of primary particle size will not be enough to predict deposition, and the formulation of particles for pulmonary delivery will require knowledge of the dynamics of aerosol behavior for efficient targeting.

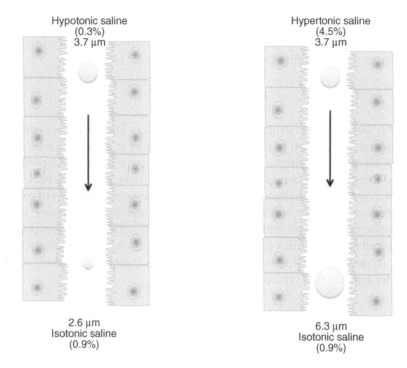

Figure 1.3 Hygroscopic growth and shrinkage of hypertonic and hypotonic droplets of the same initial size (3.7 μm) in the humid environment of the respiratory tract

1.8 Lung clearance mechanisms

Once deposited on the surface of the airways, the fate of particles will depend on their solubility and the landing site (see Table 1.2).

Inhaled particles can be dissolved in lung fluid, can act locally or pass into the systemic system, or can be translocated out of the respiratory tract when insoluble [20–22].

The mucus that lines the pulmonary epithelium (1–10 μm thick) and the surfactant that lines the alveoli (0.1–0.2 μm thick) constitute physical barriers to

Table 1.2 Fate of deposited particles in the lungs

Type of particle	Uptake or clearance
Soluble particle	Dissolution and blood circulation
Insoluble particle	Local action
or	Translocation, transcytosis, systemic or sensory-nerves uptake
	Lymphatic uptake (~500 nm)
	Clearance: macrophages and mucociliary

pulmonary absorption of drugs. In the normal lung, the rate of mucus movement varies with the airway region and is determined by the number of ciliated cells and their beat frequency. Movement is faster in the trachea than in the small airways and is affected by factors influencing ciliary functioning and the quantity and quality of mucus. If the deposited matter is fairly soluble in the body fluids, it will enter the blood circulation.

Lipophilic molecules pass easily through the airway epithelium via passive transport. Hydrophilic molecules cross via extracellular pathways, such as tight junctions, or by active transport via endocytosis and exocytosis [23]. From the submucosal region, particles are absorbed into the systemic circulation, bronchial circulation, or lymphatic system.

For relatively insoluble matter, clearance is governed mainly by mechanical removal of particles via phagocytosis by alveolar macrophages and mucociliary transport. Niven [24] identified respiratory mucus, mucociliary clearance, alveolar lining layers, alveolar epithelia, basement membranes, pulmonary enzymes, macrophages, and other cells as barriers to pulmonary absorption of biotherapeutic agents. Although the alveolar epithelium and capillary endothelium have high permeability to water, many gases, and lipophilic substances, the permeation of many hydrophilic substances of large molecular size and of ionic species is limited [25]. The molecular-weight cut-off of tight junctions for alveolar type I cells is 0.6 nm. Endothelial junctions allow passage of larger molecules (4–6 nm). On reaching the alveoli, most peptides and proteins are either degraded by proteases or removed by alveolar macrophages.

The retention half-time of solid particles in the alveolar region based on this clearance mechanism is about 70 days in rats and up to 700 days in humans [26]. Within 6–12 hours after deposition in the alveoli, virtually all of the particles are phagocytized, but it would appear that there are significant particle size-dependent influences on the effectiveness of this process [27]. The optimal particle size for phagocytosis by alveolar macrophages has been estimated at between 1 and 3 μm, with smaller particles resulting in a rate of phagocytosis that is progressively slower [28]. Those molecules and particles that are not removed by phagocytosis, such as nanoparticles in the deep lungs, where there are no macrophages, readily gain access to epithelial and interstitial sites, blood circulation, and even the lymphatic nodes. Once particles have translocated to the blood circulation, they can be distributed throughout the body. Different mechanisms have been proposed for uptake in systemic circulation and biological tissues. One involves transcytosis across the epithelium of the respiratory tract into the interstitium and access to the blood circulation either directly, via the lymphatic system, or through large transitory pores in the epithelium caused by cell injury or apoptosis [2,26,29].

Very little is known about the drug-metabolizing activities of the lung affect and how these affect the concentration and therapeutic efficacy of inhaled drugs. All metabolizing enzymes found in the liver are found to a lesser extent in the lung (numbers of CYP450 enzymes are 5–20 times lower in lung than in liver), distributed throughout the conducting airways and alveoli [30–32]. However, for most proteins, degradation in the alveoli is not a major clearance mechanism, with >95% of proteins, including insulin, being absorbed intact from the lung periphery [29,33].

1.9   Local and systemic drug delivery

The lung offers a large surface area for drug absorption [2]. In addition, the alveolar epithelium is very thin (approximately 0.1–0.5 μm thick) [34], permitting rapid drug absorption. The alveoli can be effectively targeted for drug absorption by delivering the drug as an aerosol with MMAD <5 μm. Inhalation has long been established in the treatment of both local respiratory and systemic diseases, as it is an effective means of delivering drugs for both. The advantages of inhaled local versus systemic delivery of various drugs are listed in Table 1.3.

Until recently, aerosol drug delivery was mostly limited to topical therapy for the lung and nose. The major contributing factor to this restriction was the inefficiency of available devices, which could deposit only 10–15% of the emitted dose in the lungs. While appropriate lung doses of local therapy can be achieved with these devices, for systemic therapies large amounts of drug are necessary in order to achieve therapeutic drug levels. Recent advances in aerosol and formulation technologies have led to the development of delivery systems that are more efficient and which produce small-particle aerosols, allowing higher drug doses to be

Table 1.3 Advantages of local and systemic drug delivery to the lung

Local delivery	Systemic delivery
Deliver high concentration directly to the disease site, minimizing risk of systemic side-effects	Non-invasive delivery system
Rapid clinical response	Suitable for a wide range of substances, from small molecules to large proteins [35,36]
Bypass GI absorption and first-pass metabolism in the liver	Large absorptive surface area and high permeable membrane in the alveolar region [2]
Achieve a similar or superior therapeutic effect at a fraction of the systemic dose	Less-harsh and low-enzymatic environment devoid of first-pass metabolism
	Reproducible absorption kinetics, since pulmonary delivery is independent of extracellular enzymes and metabolic differences, such as for the GI [36]

Table 1.4 Compounds delivered via the pulmonary route

	Small molecules	Large molecules
Respiratory disease	Inhaled corticosteroids [35]	Peptide agonists/antagonists [36]
	β_2 agonists [37]	Antibodies (anti-IgE) [38]
	Anticholinergics [39]	DNA (genes) (CFTR) [40]
	Antibiotics [41]	Aptamers [42]
	Antifungals [43]	
Nonrespiratory disease	Morphine [44]	Peptide agonists/antagonists [36]
	Anesthetics [45]	Antibiotics [41]
	$5HT_{1B/1D}$ agonists (triptans)[46]	DNA (genes)[47]
	Adenosine A1 agonists [48]	Aptamers [42]
	Sildenafil [49]	Vaccines [50]

deposited in the alveolar region of the lungs, where they are available for systemic absorption.

Several compounds of various molecular sizes can be delivered via the lung to treat a range of diseases, including respiratory and nonrespiratory conditions. A summary is presented in Table 1.4.

1.10 Conclusion

Although not without barriers, as described briefly in this chapter, the lung is a very desirable target for drug delivery. It not only provides direct access to the site of disease for the treatment of local respiratory diseases, without the inefficiencies and unwanted effects of systemic drug delivery, but also has an enormous surface area to be utilized for the delivery of systemic absorption of medications, and a relatively low enzymatic count. Airway geometry, humidity, clearance mechanisms, and the presence of lung disease influence the deposition of aerosols and therefore the therapeutic effectiveness of inhaled medications. To provide an efficient and effective inhalation therapy, these factors must be considered.

References

1. Weibel E. Morphometry of the Human Lung. Berlin: Springer Verlag; 1963.
2. Patton JS. Mechanisms of macromolecule absorption by the lungs. Advance Drug Delivery Reviews 1996;19:3–36.
3. Moren F, Dolovich M, Newhouse M, Newman S. Aerosols in Medicine: Principles, Diagnosis and Therapy. Amsterdam: Elsevier Science Publishers; 1993.
4. O'Rahilly R. Basic Human Anatommy. Philadelphia, PA: W.B. Saunders; 1983.

5. Smyth HDC. The influence of formulation variables on the performances of alternative propellant-driven metered dose inhalers. Advance Drug Delivery Reviews 2003;55:807–828.

6. Chrystyn H. Is total particle dose more important than particle distribution? Respiratory Medicine 1997;91:17–19.

7. Zanen P, Go LT, Lammers JWJ. The optimal particle-size for beta-adrenergic aerosols in mild asthmatics. International Journal of Pharmaceutics 1994;107: 211–217.

8. Heyder J, Gebhart J, Rudolf G, Schiller CF, Stahlhofen W. Deposition of particles in the human respiratory-tract in the size range 0.005-15-Mu-M. Journal of Aerosol Science 1986;17:811–825.

9. Hinds WC. Aerosol Technology. New York, NY: John Wiley & Sons, Ltd; 1999.

10. Patton JS, Byron PR. Inhaling medicines: delivering drugs to the body through the lungs. Nature Reviews Drug Discovery 2007;6:67–74.

11. Clarke SW. Medical aerosol inhalers: past present and future. In Clarke SW, Pavia D, editors. *Aerosola and the Lung: Clinical and Experimental Aspects*. London: Butterworths; 1984. pp. 1–18.

12. Lourenco RV, Cotromanes E. Clinical aerosols. 1. Characterization of aerosols and their diagnostic uses. Archives of Internal Medicine 1982;142:2163–2172.

13. Heyder J. Particle-transport onto human airway surfaces. European Journal of Respiratory Diseases 1982;63:29–50.

14. Kim CS, Jaques PA. Analysis of total respiratory deposition of inhaled ultrafine particles in adult subjects as various breathing patterns. Aerosol Science and Technology 2004;38:525–540.

15. Smith S, Cheng US, Yeh HC. Deposition of ultrafine particles in human tracheo-bronchial airways of adults and children. Aerosol Science and Technology 2001;35:697–709.

16. International Commission on Radiological, Protection. Human Respiratory Tract Model for Radiological Protection. Publication 66. Oxford: Elsevier; 1994.

17. Newman SP. Aerosol deposition considerations in inhalation-therapy. Chest 1985;88:S152–S160.

18. Courrier H, Butz N, Vandamme T. Pulmonary drug delivery systems: recent developments and prospects. Critical Reviews in Therapeutic Drug Carrier Systems 2002;19:425–498.

19. Phipps PR, Gonda I, Anderson SD, Bailey D, Bautovich G. Regional deposition of saline aerosols of different tonicities in normal and asthmatic subjects. European Respiratory Journal 1994;7:1474–1482.

20. Martonen T. On the fate of inhaled particles in the human—a comparison of experimental-data with theoretical computations based on a symmetric and asymmetric lung. Bulletin of Mathematical Biology 1983;45:409–424.

21. Blank F, Rothen-Rutishauser BM, Schurch S, Gehr P. An optimized in vitro model of the respiratory tract wall to study particle cell interactions. Journal of Aerosol Medicine—Deposition Clearance and Effects in the Lung 2006;19:392–405.

22. Finlay WH. The Mechanics of Inhaled Pharmaceutical Aerosols: An Introduction. London: Academic Press; 2001.

23. Summers QA. Inhaled drugs and the lung. Clinical and Experimental Allergy 1991;21:259–268.

24. Niven RW. Delivery of biotherapeutics by inhalation aerosol. Critical Reviews in Therapeutic Drug Carrier Systems 1995;12:151–231.

25. Sayani AP, Chien YW. Systemic delivery of peptides and proteins across absorptive mucosae. Critical Reviews in Therapeutic Drug Carrier Systems 1996;13:85–184.

26. Oberdorster G, Oberdorster E, Oberdorster J. Nanotoxicology: an emerging discipline evolving from studies of ultrafine particles. Environmental Health Perspectives 2005;113:823–839.

27. Lehnert BE. Pulmonary and thoracic macrophage subpopulations and clearance of particles from the lung. Environmental Health Perspectives 1992;97:17–46.

28. Oberdorster G, Ferin J, Lehnert BE. Correlation between particle-size, in-vivo particle persistence, and lung injury. Environmental Health Perspectives 1994; 102:173–179.

29. Folkesson HG, Matthay MA, Westrom BR, Kim KJ, Karlsson BW, Hastings RH. Alveolar epithelial clearance of protein. Journal of Applied Physiology 1996; 80:1431–1445.

30. Upton RN, Doolette DJ. Kinetic aspects of drug disposition in the lungs. Clinical and Experimental Pharmacology and Physiology 1999;26:381–391.

31. Krishna DR, Klotz U. Extrahepatic metabolism of drugs in humans. Clinical Pharmacokinetics 1994;26:144–160.

32. Dahl AR, Lewis JL. Respiratory-tract uptake of inhalants and metabolism of xenobiotics. Annual Review of Pharmacology and Toxicology 1993;33: 383–407.

33. Hastings RH, Grady M, Sakuma T, Matthay MA. Clearance of different-sized proteins from the alveolar space in humans and rabbits. Journal of Applied Physiology 1992;73:1310–1316.

34. Wilson C, Washington N. Physiological pharmaceutics. Chichester: John Wiley & Sons, Ltd; 1989.

35. Byron PR, Patton JS. Drug-delivery via the respiratory-tract. Journal of Aerosol Medicine—Deposition Clearance and Effects in the Lung 1994;7(1):49–75.

36. Skoner DP, Angelini BL, Friday G, Gentile D. Clinical use of nebulized budesonide inhalation suspension in a child with asthma. Journal of Allergy and Clinical Immunology 1999;104(4):S210–S214.

37. Laube BL, Benedict GW, Dobs AS. Time to peak insulin level, relative bioavailability, and effect of site of deposition of nebulized insulin in patients with noninsulin-dependent diabetes mellitus. Journal of Aerosol Medicine—Deposition Clearance and Effects in the Lung 1998;11(3):153–173.

38. Becker AB, Simons FE. Comparison of formoterol, a new long-acting beta-agonist, with salbutamol and placebo in children with asthma. Journal of Allergy and Clinical Immunology 1989;84(1):185.

39. Fahy JV, Cockcroft DW, Boulet L-P, Wong HH, Deschesnes F, Davis EE, Ruppel J, Su JQ, Adelman DC. Effect of aerosolized anti-IgE (E25) on airway responses to inhaled allergen in asthmatic subjects. American Journal of Respiratory and Critical Care Medicine 1999;160(3):1023–1027.

40. ZuWallack AR, ZuWallack RL. Tiotropium bromide, a new, once-daily inhaled anticholinergic bronchodilator for chronic-obstructive pulmonary disease. Expert Opinion on Pharmacotherapy 2004;5(8):1827–1835.

41. Alton, EW, Stem M, Farley R, Jaffe A, Chadwick SL, Phillips J, Davies J, Smith SN, Browning J, Davies MG, Hodgson ME, Durham SR, Li D, Jeffery PK, Scallan M, Balfour R, Eastman SJ, Cheng SH, Smith AE, Meeker D, Geddes DM. Cationic lipid-mediated CFTR gene transfer to the lungs and nose of patients with cystic fibrosis: a double-blind placebo-controlled trial. Lancet 1999;353(9157):947–954.

42. Adi H, Young PM, Chan H-K, Stewart P, Agus H, Traini D. Cospray dried antibiotics for dry powder lung delivery. Journal of Pharmaceutical Sciences 2008;97(8):3356–3366.

43. Eckstein F. The versatility of oligonucleotides as potential therapeutics. Expert Opinion on Biological Therapy 2007;7(7):1021–1034.

44. Vaughn JM, Wiederhold NP. Murine airway histology and intracellular uptake of inhaled amorphous itraconazole. International Journal of Pharmaceutics 2007;338 (1–2):219–224.

45. Mallet JP, Diot P, Lemarie E. Aerosols for administration of systemic drugs. Revue Des Maladies Respiratoires 1997;14(4):257–267.

46. Hatch DJ. New inhalation agents in paediatric anaesthesia. British Journal of Anaesthesia 1999;83(1):42–49.

47. Dahlof C. Sumatriptan nasal spray in the acute treatment of migraine: a review of clinical studies. Cephalalgia 1999;19(9):769–778.

48. Harvey B-G, Hackett NR, Crystal RG. Airway epithelial CFTR mRNA expression in cystic fibrosis patients after repetitive administration of a recombinant adeno-virus. Journal of Clinical Investigation 1999;104(9):1245–1255.

49. Russo C, Arcidiacono G, Polosa R. Adenosine receptors: promising targets for the development of novel therapeutics and diagnostics for asthma. Fundamental & Clinical Pharmacology 2006;20(1):9–19.

50. Ghofrani HA, Osterloh IH, Grimminger F. Sildenafil: from angina to erectile dysfunction to pulmonary hypertension and beyond. Nature Reviews Drug Discovery 2006;5(8):689–702.

51. LiCalsi C, Christensen T, Bennett JV, Phillips E, Witham C. Dry powder inhalation as a potential delivery method for vaccines. Vaccine 1999;17(13–14):1796–1803.

2

Inhalation and nasal products

Daniela Traini and Paul M. Young

Respiratory Technology, The Woolcock Institute of Medical Research & The Discipline of Pharmacology, The University of Sydney, Sydney, Australia

2.1 Introduction

Pharmacologically active substances have been administered to humans for thousands of years, but delivery via the lung is a relatively recent development. Commercialization of inhaled respiratory medicines was not achieved until 1948, when Abbot Laboratories developed the Aerohaler for inhaled penicillin G powder; they then revolutionized the field in 1955 with the advent of the pressurized metered-dose inhaler (pMDI) [1]. Since this inception, the range of inhaler products and medicaments has grown to encompass alternative drug delivery systems, namely those based on the dry powder inhaler (DPI), nebulizers, and nasal products. This chapter will review the three main inhalation delivery platforms (DPIs, pMDIs, and nebulizers), as well as nasal drug-delivery methods and formulations.

2.2 Dry powder inhalers (DPIs)

DPIs, as their name implies, contain and deliver the active medicament as a dry powder of suitable aerodynamic size for respiratory therapy. Dry powder particles of a suitable size, generally less than 6 microns [2], can be readily produced by micronization or by other size-reduction methods. However, such particles have high surface area to mass ratios, making them highly cohesive/adhesive. Consequently, the drug must be formulated in such a way that the energy input during inhalation is sufficient to overcome the contiguous adhesive and cohesive particle forces, aerosolizing the powder for respiratory deposition. Although in principle this

Inhalation Drug Delivery: Techniques and Products, First Edition. Paolo Colombo, Daniela Traini, and Francesca Buttini.
© 2013 John Wiley & Sons, Ltd. Published 2013 by John Wiley & Sons, Ltd.

approach may appear straightforward, the physicochemical nature and interactive mechanisms of the components in a DPI system are still relatively poorly understood. Many commercial products have, by pharmaceutical standards, relatively poor efficiencies, with often less than 20% drug being delivered to the lung [3]. This has generated a significant amount of research activity in the fields of pharmaceutics, powder technology, and surface, aerosol, and colloid science, which has focused on understanding the interactions in DPI systems.

The specific technologies underpinning DPI formulation are discussed in Chapter 3, but basically the efficacy of a DPI formulation is dependent on both the formulation of the powder and the design of the device. There are two main approaches to formulating the drug powder:

1. carrier-based formulations containing larger inert carrier excipients;

2. agglomerated drug-only or binary systems.

Examples of carrier-based systems include GlaxoSmithKline's (GSK's) Ventolin Diskus and Novartis's Foradil Aeroliser products. Examples of agglomerated drug-only formulations include Astra Zeneca's (AZ's) Pulmicort and Bricanil Turbohalers. A company product will come with a formulation (drug name) and a device name. Some formulations are reformulated in multiple devices (for example, GSK's Ventolin has been available in both the Diskhaler and Diskus devices).

In terms of design, DPI devices are either single-dose systems (which require refilling by the patient) or multi-dose systems (where the patient activates the dose pre-inhalation). Single-dose systems usually utilize capsules to store the formulation, which are inserted into the device, and pierced by the patient, prior to inhalation. Multi-dose systems store the powder either in individual blisters (which are fired sequentially) or in a powder reservoir (which is metered by the device). A chart outlining the different forms of device, along with examples, is given in Figure 2.1.

An important thing to note with commercially available DPIs is that they are all passive devices; passive devices rely on the patients inspiratory flow rate to generate the energy for powder dispersion. It is therefore important for inhaler developers to design devices that (1) allow powder dispersion upon inhalation at reasonable flow rates and (2) have flow-rate independence. It is important to note that different patients will have different lung capacities and maximum ventilatory flows (according to their demographic and disease severity), and thus variation in device performance and resistance with respect to flow is critical to consistent efficacy [4].

DPI devices have been around for more than 20 years and thus a wide range of devices have been or are currently available on the market. A collection of these devices is shown in Figure 2.2.

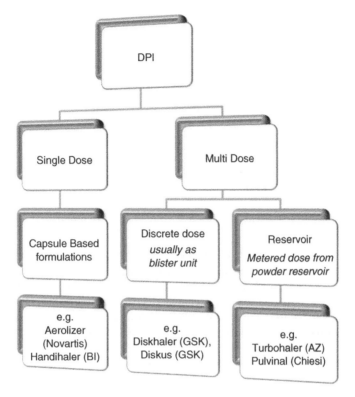

Figure 2.1 DPI-device approaches and product examples

Another important issue to consider regarding DPI devices is their ease of use. The evolution of DPI devices has come about, in part, due to coordination issues associated with their pMDI counterparts. However, while DPI devices are relatively free from coordination issues, device complexity must be considered when prescribing medicines. Many of the first-generation and some of the second-generation DPI devices require multiple steps to ensure a dose is loaded and/or may require specific cleaning routines after use (Table 2.1 shows the Relenza operating instructions, for example). The former point must be considered when selecting a device for patients who have potential dexterity issues. In addition, when introducing new or alternative devices to a patient, which have open reservoirs or are moisture-sensitive, storage (in a bathroom cabinet, for example) must be discussed in order to avoid degradation of the formulation. It is important to ensure that the prescription of a new device and formulation is accompanied by adequate education from the general practitioner, consultant, and pharmacist.

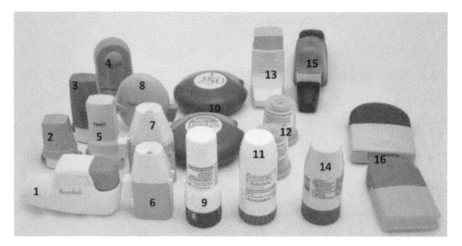

Figure 2.2   Examples of past and present DPI devices. (1) Novolizer by Almirall S. A.,
(2) Aerolizer. The AEROLIZER® photo image is reproduced with permission of Merck Sharp &
Dohme Corp., subsidiary of Merck & Co., Inc. All rights reserved. The trademark AEROLIZER® is a
registered trademark of Novartis AG., (3) Cyclohaler Courtesy of Teva UK Ltd., (4) Spinhaler by
Rhodia (Fisons), (5) Aerolizer. The AEROLIZER® photo image is reproduced with permission of
Merck Sharp & Dohme Corp., subsidiary of Merck & Co., Inc. All rights reserved. The trademark
AEROLIZER® is a registered trademark of Novartis AG., (6) Inhalateur Ingelheim. Courtesy of
Boehringer Ingelheim Ltd., (7) Aerohaler/Inhalator. Courtesy of Boehringer Ingelheim Ltd.,
(8) Handihaler. Courtesy of Boehringer Ingelheim Ltd., (9) Twisthaler. The TWISTHALER® photo
image is reproduced with permission of Merck Sharp & Dohme Corp, subsidiary of Merck & Co., Inc.
All rights reserved. The trademark TWISTHALER® is a registered trademark of Merck Sharp &
Dohme Corp., (10) Accuhaler/Diskus by GlaxoSmithKline Respiratory, (11) Turbohaler. Courtesy of
AstraZeneca UK Ltd., (12) Auto-Jethaler/Auto-haler by RatioPharm (PulmoTec), (13) Clickhaler.
Courtesy of Innovata Biomed Ltd., (14) Pulvinal. Courtesy of Chiesi Farmaceutici, Parma, Italy, (15)
Easyhaler by Orion Pharma, (16) Diskhaler by GlaxoSmithKline Respiratory (Allen & Hanburys)

2.3   Liquid and propellant-based inhalers

Liquid and propellant aerosol systems can be generally subclassified into nebulizer and
pMDI technologies. The main difference between the two is that nebulizers utilize an
external energy source to produce aerosolized fine particulate droplets of the formulation while pMDI systems incorporate a propellant into the formulation, which provides
the energy for aerosolization upon actuation. Both of these approaches have advantages
and disadvantages, both can contain either solution- or suspension-based formulation strategies, and both have undergone technological advances in recent years.

2.3.1   Pressurized metered-dose inhalers (pMDIs)

pMDIs (or MDIs in the USA) were introduced in the 1950s by Riker Laboratories [5].
Initially the formulation was contained in a chlorofluorocarbon (CFC) propellant, but

Table 2.1 Patient instruction leaflet for the Diskhaler Relenza DPI inhaler [5]. RS01 DPI - Plastiape. Courtesy of Plastiape IT

Step A: Load the medicine into the DISKHALER	Step B: Puncture the blister	Step C: Inhale	Step D: Move the medicine disk to the next blister
1 Start by pulling off the blue cover. 2 Always check inside the mouthpiece to make sure it is clear before each use. If foreign objects are in the mouthpiece, they could be inhaled and cause serious harm. 3 Pull the white mouthpiece by the edges to extend the white tray all the way. 4 Once the white tray is extended all the way, find the raised ridges on each side of it. Press in these ridges, both sides at the same time, and pull the whole white tray out of the DISKHALER body. 5 Place one silver medicine disk on to the dark brown wheel, flat side up. The four silver blisters on the underside of the medicine disk will drop neatly into the four holes in the wheel. 6 Push in the white tray as far as it will go. Now the DISKHALER is loaded with medicine.	The DISKHALER punctures one blister of medicine at a time so you can inhale the right amount. It does not matter which blister you start with. Check to make sure that the silver foil is unbroken. 1 Be sure to keep the DISKHALER level so the medicine does not spill out. 2 Locate the half-circle flap with the name "RELENZA" on top of the DISKHALER. 3 Lift this flap from the outer edge until it cannot go any farther. Flap must be straight up for the plastic needle to puncture both the top and bottom of the silver medicine disk inside. 4 Keeping the DISKHALER level, click the flap down into place.	1 Before putting the white mouthpiece into your mouth, breathe all the way out (exhale). Then put the white mouthpiece into your mouth. Be sure to keep the DISKHALER level so the medicine does not spill out. 2 Close your lips firmly around the mouthpiece. Be sure not to cover the small holes on either side of it. 3 Breathe in through your mouth steadily and as deeply as you can. Your breath pulls the medicine into your airways and lungs. 4 Hold your breath for a few seconds to help RELENZA stay in your lungs where it can work. To take another inhalation, move to the next blister by following Step D. Once you've inhaled the number of blisters prescribed by your health care provider, replace the cover until your next dose.	1 Pull the mouthpiece to extend the white tray, without removing it. 2 Then push it back until it clicks. This pull–push motion rotates the medicine disk to the next blister. 3 To take your next inhalation, repeat Steps B and C. If all four blisters in the medicine disk have been used, you are ready to start a new medicine disk (see Step A). Check to make sure that the silver foil is unbroken each time you are ready to puncture the next blister.

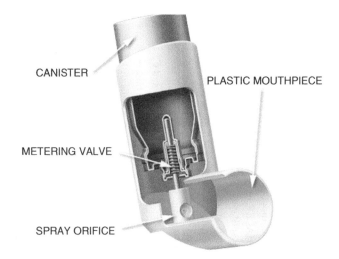

CANISTER

PLASTIC MOUTHPIECE

METERING VALVE

SPRAY ORIFICE

Figure 2.3 Schematic of the major components of a pMDI. Courtesy of 3M Drug Delivery Systems

this was superseded by hydrofluoroalkane (HFA) propellants (between the 1990s and early 2000s) due to the Montreal Protocol ban on the former, ozone-damaging molecules [6]. The canister is typically a formed single aluminum component with a metering valve, crimped to the canister lid. The metering valve isolates the pressurized formulation's local environment (typically 4–5 atmospheres of pressure) from the surrounding environment (1 atmosphere), and upon actuation releases a known quantity of propellant formulation (typically 25–100 μL) to the surrounding environment. The formulation is either solubilized or suspended in the propellant, and the local concentration of drug in the canister determines the therapeutic dose. Expansion of the propellant as it exits the spray orifice results in a fine mist of HFA, which evaporates off to leave a micron-sized drug system suitable for inhalation. A schematic of the main components of a pMDI is shown in Figure 2.3.

The advantages of pMDI-based inhalation systems are their portable nature, ease of operation, and familiarity (although this may not necessarily be an advantage, since familiarity does not necessarily go hand-in-hand with correct use). In terms of disadvantages, pMDI devices have a smaller delivery volume than their DPI and nebulizer counterparts (limiting the range of therapeutic doses). Furthermore, some patients complain of a cooling "Freon" effect, due to the rapid evaporation of the propellant [7]. However, the most important patient-related issue with respect to these devices is the requirement for a coordinated inhalation and actuation maneuver. The pMDI is an active device, delivering a single dose over a period of microseconds. In comparison to nebulizers (which patients inhale over minutes of tidal breathing) or DPIs (which rely on the patient's inspiratory flow to aerosolize the powder), the inhalation of pMDIs requires a coordinated effort during actuation.

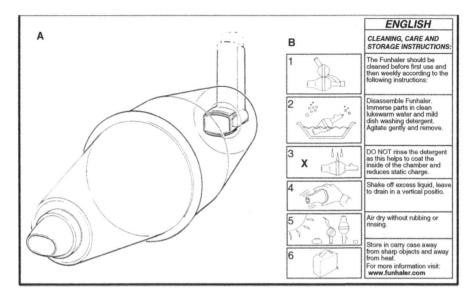

Figure 2.4 (a) Schematic of the GSK Volumatic Spacer device by GlaxoSmithKline. (b) Instruction leaflet for the Avita Funhaler. Image modified from Rotherham General Hospitals NHS Trust. Courtesy of Avita Medical Ltd

If the patient inhales before or after actuation, the majority of the medicament is likely to impact in the throat and be swallowed.

To overcome such issues, two approaches have been taken: spacers and breath-actuated pMDIs. Both attempt to overcome issues with poor patient coordination and have advantages and disadvantages.

Spacers are add-ons which attach to the mouthpiece of the pMDI. The patient or care worker actuates the device, delivering the aerosol into the spacer, so that the patient can inhale via tidal breathing (Figure 2.4a).

Since spacers are add-on devices, little or no modification to the pMDI needs to take place. Furthermore, the use of these devices via tidal breathing allows a therapeutic dose to be received over multiple breaths; this is especially useful in young and elderly patients. Spacers are produced as device-specific add-on units (e.g. GSK's Volumatic Spacer or 3M's Aerochamber) or as generic units for use with multiple pMDI medicines (e.g. Avita's Funhaler).

The addition of a spacer add-on to a pMDI significantly increases the overall device size. While in many cases this may not be an issue, portability should be considered. Furthermore, since the spacer is an intermediate "volume" between the pMDI and the patient, variations in the delivered dose and aerosol cloud properties will occur when compared to the pMDI alone [8].

It is also important to consider care and handling. Most spacer devices have specific instructions regarding cleaning (Figure 2.4b). Generally, it is recommended that spacer units are cleaned with a detergent or soapy water and are not rinsed or cloth-dried. This is due to the potential static-charge build-up on the inner surfaces of the spacer walls, which has been reported to cause aerosol particle attraction and variation in dose performance [9].

Breath-actuated pMDIs usually contain modified casings, which, upon actuation, hold the canister in a primed and post-actuated condition. The patient inhales through the device and the dose is released, removing the requirement for coordination. Breath-actuated systems are of a similar size to their regular pMDI counterparts and require similar inhalation maneuvers (i.e. forced inspiratory rather than tidal breathing).

One further limitation of pMDI technology has been the lack of dose counters. With capsule-based DPIs, the patient can physically see how many doses are available, while unit-dose and reservoir DPI systems routinely include dose counters. Historically, patients were told to test how many doses remained in a pMDI by submerging the canister in a bowl of water and estimating the number of doses remaining via buoyancy. This is no longer recommended, since (1) it is inaccurate and (2) it may cause potential valve swelling, water ingress, and formulation damage. Recently, dose counters have been included in pMDI formulations, circumventing these issues [10].

Ultimately, the choice of pMDI and add-on device should be based on the drug used (e.g. a once-a-day steroid versus a "carry-everywhere" reliever) and the patient's requirements (including age, frailty, severity of disease, and need for convenience).

2.3.2 Nebulizers

Nebulizers are designed to produce a constant stream of aerosol particles at a suitable size for respiratory delivery. Conventional nebulizers utilize compressed air to generate a fine aerosol mist that can be inhaled over sustained periods, via tidal breathing (however, other designs—such as vibrating-mesh and ultrasonic designs—do exist). Jet nebulizers contain two main components: (1) the nebulizer (containing mouthpiece, drug reservoir, and aerosolization mechanism) and (2) the compressor (Figure 2.5). Historically, the compressor unit was relatively large (with a footprint approximately A4 in size) and expensive (although cheaper units are available today). The nebulizer component can be either reusable or disposable and the medication is usually supplied in the form of ampoules, which contain a sterile solution of the medicament required. Due to the relative size and expense of nebulizers, they have generally been used in the hospital setting, or at home for severe respiratory conditions where higher doses or prolonged treatment are required. For comparison, the Ventolin pMDI formulations contain 100 µg salbutamol per dose; their nebulizer equivalent might contain up to 5000 µg per treatment.

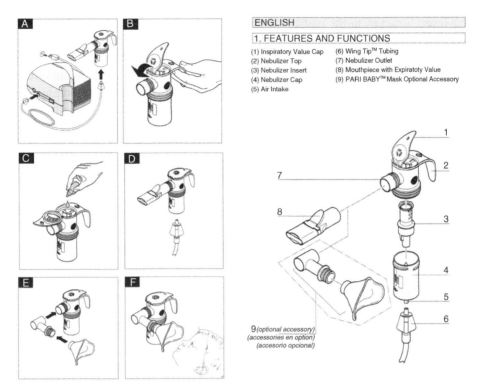

Figure 2.5 Features and assembly of the PARI LC Sprint reusable nebulizer. Courtesy of PARI Respiratory Equipment Inc.

Since nebulizers produce a constant aerosol stream, there is a potential risk that health care workers and family members will be exposed to "second-hand" aerosol during patient exhalation [11]; however, these risks may be reduced via clever engineering of the nebulizer design (e.g. in the breath-enhanced nebulizers by PARI Ltd and breath-actuated AeroEclipse nebulizer by Monaghan Medical Corp). Furthermore, developments in vibrating-mesh technologies (e.g. in the PARIs eFLOW and Respironics I-neb) have led to miniaturization of nebulizer design, resulting in handheld units that, in future, may be used to deliver equivalent doses to their jet-nebulizer counterparts in a more efficient and convenient manner.

2.4 Nasal formulations

Nasal drug products are found as both prescription and over-the-counter medicines. They are used to treat a diverse range of common diseases, such as the cold and hayfever; however, in the past two decades they have found a niche market, as an alternative to parenteral injections. The nasal cavity is highly vascularized and offers a significant advantage as a delivery route for systemically delivered medicines (specifically, the lack of requirement for needle use). In addition, this route offers advantages in terms of fast action, improved bioavailability, and patient compliance. There have therefore been a growing number of marketed systemic medicines.

In addition to topical and systemic delivery routes, nasal formulations also offer a potential means by which to circumvent the blood–brain barrier, allowing direct targeting of the brain. While this is still in the early stages of development, such an advance would allow the treatment of common intracerebral diseases such as Parkinson's and Alzheimer's.

There are clearly many advantages to nasal delivery, but disadvantages and significant challenges exist. These include potential local tissue irritation, rapid removal of the delivery site via clearance, and variation in the adsorption profile due to physiology and to pathologic conditions such as cold or allergy.

2.4.1 Nasal physiology

The nasal cavity contains a right and a left side, separated by a section of bone and cartilage called the septum. The internal cavity surfaces are lined with a mucosa, which contains blood vessels, glands, and cilia. The glands produce a mucus coating that is transported by the cilia to the back of the nasal cavity, where the mucous is swallowed. This movement of mucus by the cilia is termed the "mucociliary clearance mechanism" and provides an efficient "defense" system for capturing bacteria and airborne particulates for removal and digestion. Ultimately, this protects the respiratory system. The cilia are cellular protrusions. They beat with a frequency of <1000 strokes per minute and transport mucus at a rate of approximately 5 mm per minute. It has been reported that the mucociliary clearance mechanism is capable of clearing objects from the nasal cavity with a half-life of around 15 minutes [12].

Assuming the nasally-delivered drug is not cleared from the cavity by the cilia, it can cross the mucosa via two different pathways: transcellularly (across the cell) or paracellularly (between cells). In general, lipophilic drugs are transported transcellularly via passive diffusion or by receptor or vesicular transport, while polar drugs are transported paracellulary, passing between the cells via the tight junctions (although active transport may play a role in some cases). Since the tight junctions are small (<10 nm) [13], the paracellular route is limited by the molecular size of the drug. This route is thus less efficient for large molecules/macromolecules. In general, a size cut-off of <1000 Da is considered the upper limit for paracellular drug delivery [13]. Lipophilic drugs (e.g. propanolol and progesterone) have shown rapid absorption profiles when given nasally. Indeed, the bioavailability of such molecules has approached 100% for some drugs [14]. In comparison, the bioavailability of polar compounds via nasal absorption is poor, with a bioavailability of <10% for small molecules such as morphine and <1% for macromolecules such as insulin. For polar drugs, in the absence of an active transport system, the mucociliary clearance mechanism has a significant effect on bioavailability.

2.4.2   Delivery issues and concerns

The deposition site of a drug formulation in the nasal cavity is important, since different regions have differing permeabilities and cilia concentrations. The deposition profile of a formulation, however, will be dependent on the formulation characteristics (particle size and velocity), the mode of administration, and the internal geometry of the nasal cavity.

The particle size of a nasal formulation is not considered as critical as it is for an inhalation formulation. This is partly due to the fact that the formulation is delivered at close proximity to the target site. However, since the nasal mechanism is designed to capture particulate matter, so long as the material is not respirable in nature, the target will be met. In general, a particle size between 5 and 10 microns is considered ideal.

Most nasal formulations are prepared as liquids and delivered by a conventional press-down spray pump. The size distributions are thus not particularly narrow. Due to the nature of this delivery system, a significant component of the bolus dose is deposited on the anterior segment lined by skin, which is a target for neither topical nor systemic drugs. Such deposition may result in adverse physiological effects, including, in some cases, irritation and bleeding. Finally, inconsistent deposition in target-specific regions (i.e. sinuses, middle ear, and the olfactory region) using current technology makes the delivery of systemic drugs and drugs targeting the blood–brain barrier difficult.

A summary of factors affecting nasal drug delivery is given in Table 2.2.

Table 2.2   Factors affecting nasal delivery

Drug physicochemical properties	Nasal physiology	Delivery system
Molecular size	Membrane permeability	Concentration
Lipophilic–hydrophilic balance	Environmental pH	pH
Enzymatic degradation	Mucociliary clearance	Osmolarity
	Mucosa irritation	Drug distribution and deposition
		Effect on epithelial membrane

2.4.3   Strategy for enhanced nasal delivery

Nasal absorption may be improved via penetration enhancers. Penetration enhancers are active excipients that promote the transport of a drug across the nasal mucosa [15]. In addition, excipients which enhance the residence time in the nasal cavity (bioadhesive molecules) also have the potential to improve drug transport and bioavailability.

There are many penetration enhancers (including chelators, fatty acid salts, phospholipids, glycos, and cyclodextrins), and their action can be by one or a combination of the following mechanisms:

1. alteration of properties of the mucosa layer;

2. opening of tight junctions between epithelial cells;

3. reversed formation of micelles between membranes;

4. increase of membrane fluidity.

Interestingly, in the case of aqueous nasal formulations, the role of the preservative should be evaluated not only with respect to efficacy, but also with an eye on safety, given that some preservatives have been proved to affect the permeability characteristics of the nasal mucosa [16].

2.4.4   Marketed nasal products

In the past few years, a number of systemic drugs have been developed or marketed as nasal products. Examples are listed in Table 2.3.

Clearly, the nose is a delivery route worth considering for many existing substances, as an alternative to the intravenous route. Progress in nasal formulation and delivery technologies may offer essential advantages and expand the market for simple, painless, and long-term nasal delivery of drugs and vaccines.

2.4.5   Pharmaceutical development studies for nasal products

There are a wide variety of nasal products, and depending on the formulation type, specific pharmacopeia tests are required (Table 2.4). These tests are conducted during product development to ensure that the formulation and packaging meet the required robustness in terms of performance and stability [17]. As can be seen from Table 2.4, not all tests are required for every nasal product. It is important to note that for nasal products, there is no requirement to assess fine particle mass, since no

Table 2.3 Examples of marketed nasal products

Active ingredient	Company	Therapeutic indication
Sumatriptan	GlaxoSmithKline	Migraine
Zolmitriptan	AstraZeneca	Migraine
Ergotamine	Novartis	Migraine
Butorphanol	Bristol-Mayer Squibb	Migraine
Estradiol	Servier	Menopause
Desmopressin	Ferring	Primary nocturnal enuresis
		Withdrawn for toxicological issues
Buserelin	Aventis	Endometriosis
Calcitonin	Novartis	Osteoporosis
Antigen-adjuvant system	Berna Biotech	Influenza vaccine
		Withdrawn for toxicological issues
Cold-adapted virus system	Aviron	Influenza vaccine
Apomorphine	Britannia	Erectile dysfunction
		Parkinson's
Sumatriptan	Optinose	Migraine
Insulin	NanoDerma	Diabetes
Morphine	Nastech	Pain relief
Arginine-vasopressin	DelSite, Inc.	Control of urine output in patients with diabetes insipidus

distinction is made at this level between emitted dose and respirable dose. In fact, for nasal therapy, the dose is directly delivered into the nasal cavity, where it must impact and deposit on the mucosa, without further travelling into the airstream and reaching sites away from the nose. However, with the exception of nasal drops, particle/droplet size distribution has to be evaluated for all nasal products, because a particle size greater than 20 microns should be maintained in order to minimize deposition in the lung and GI tract (risk of bioavailability changes) [18,19].

The FDA guidance document for nasal products suggests size distribution, and its reproducibility, as well as the dynamics of spray production, should be used to characterise in vitro bioavailability and bioequivalency for locally acting drugs delivered by nasal aerosol or nasal spray [20]. Such measurements may be carried out using in-line laser-diffraction techniques (e.g. the Spraytech (Malvern Instruments, Malvern, UK) or equivalent).

2.5 Conclusion

It is evident that many inhalation and nasal products exist for the treatment of a wide variety of disease states (both local and systemic). These products utilize a variety of formulation and device designs in order to ensure product reproducibility, stability,

Table 2.4 Pharmaceutical development studies for nasal products

Pharmaceutical development study	Pressurized metered-dose nasal sprays	Nasal powders (device-metered)	Nasal liquids			
			Single-use drops	Multi-use drops	Single-use sprays	Nonpressurized multi-use metered-dose sprays
Physical characterization	Yes	Yes	Yes	Yes	Yes	Yes
Minimum fill justification	Yes	Yes	Yes	Yes	Yes	Yes
Extractable/leachables	Yes	No	Yes	Yes	Yes	Yes
Delivered-dose uniformity through container life	Yes	Yes	No	No	No	Yes
Particle/droplet size distribution	Yes	Yes	No	No	Yes	Yes
Actuator/mouthpiece deposition	Yes	Yes	Yes	Yes	Yes	Yes
Shaking requirements	Yes	No	No	No	Yes	Yes
Initial and re-priming requirements	Yes	No	No	No	Yes	Yes
Cleaning requirements	Yes	Yes	No	Yes	Yes	Yes
Low-temperature performance	Yes	No	No	No	No	No
Performance after temperature-cycling	Yes	No	No	No	Yes	Yes
Effect of environmental moisture	Yes	Yes	No	No	No	No
Robustness	Yes	Yes	Yes	Yes	Yes	Yes
Delivery-device development	Yes	Yes	Yes	Yes	Yes	Yes
Preservative effectiveness/efficacy	No	No	Yes	Yes	Yes	Yes

and performance. The formulation of DPI and pMDI medicines is discussed in more detail in Chapter 3.

References

1. Versteeg HK, Hargrave GK. Fundamentals and resilience of the original MDI actuator design. In Dalby RN, Byron PR, Peart J, Suman JD, Farr SJ, editors. Respiratory Drug Delivery X. Boca Raton, FL: Davis Healthcare International; 2006.
2. Patton JS, Byron PR. Inhaling medicines: delivering drugs to the body through the lungs. Nature Reviews Drug Discovery 2007;6(1):67–74.
3. Smith IJ, Parry-Billings M. The inhalers of the future? A review of dry powder devices on the market today. Pulmonary Pharmacology & Therapeutics 2003;16 (2):79–95.
4. Prime D, Atkins PJ, Slater A, Sumby B. Review of dry powder inhalers. Advanced Drug Delivery Reviews 1997;26(1):51–58.
5. Freedman T. Medihaler therapy for bronchial asthma: a new type of aerosol therapy. Postgraduate Medical Journal 1956;20:667–673.
6. United Nations Environment, Programme., Handbook for International Treaties for the Protection of the Ozone Layer. Kenya: United Nations Environment Programme; 1996.
7. Purewal TS, Grant DJ. Metered Dose Inhaler Technology. London: Informa Healthcare; 1997.
8. Chew NY, Chan HK. The effect of spacers on the delivery of metered dose aerosols of nedocromil sodium and disodium cromoglycate. International Journal of Pharmaceutics 2000;200(1):87–92.
9. Anhoj J, Bisgaard H, Lipworth BJ. Effect of electrostatic charge in plastic spacers on the lung delivery of HFA-salbutamol in children. British Journal of Clinical Pharmacology 1999;47(3):333–336.
10. West JJ, O'Brien O, Ellis JC, O'Brien CA. In vitro comparison of the performance of salmeterol/fluticasone propionate HFA metered dose inhaler (MDI) with and without MDI counter. Journal of Allergy and Clinical Immunology 2005;115(2): S149.
11. Tang JW, Li Y, Eames I, Chan PKS, Ridgway GL. Factors involved in the aerosol transmission of infection and control of ventilation in healthcare premises. Journal of Hospital Infection 2006;64(2):100–114.
12. Soane RJ, Frier M, Perkins AC, Jones NS, Davis SS, Illum L. Evaluation of the clearance characteristics of bioadhesive systems in humans. International Journal of Pharmeutics 1999;178:55–65.
13. McMartin C, Hutchinson LEF, Hyde R, Peters GE. Analysis of structural requirements for the absorption of drugs and macromolecules from the nasal cavity. Journal of Pharmaceutical Sciences 1987;76:535–540.

14. Illum L. Nasal drug delivery—possibilities, problems and solutions. Journal of Controlled Release 2003;1–3(87):187–198.

15. Davis SS, Illum L. Absorption enhancers for nasal drug delivery. Clinical Pharmacokinetics 2003;42(13):1107–1128.

16. Bortolotti F, Balducci AG, Sonvico F, Russo P, Colombo G. In vitro permeation of desmopressin across rabbit nasal mucosa from liquid nasal sprays: the enhancing effect of potassium sorbate. European Journal of Pharmaceutical Sciences 2009;37 (11):36–42.

17. EMEA. Guideline on the pharmaceutical quality of inhalation and nasal products. London: European Medicines Agency; 2005.

18. DeAscentiis A, Bettini R, Caponetti G, Catellani PL, Peracchia MT, Santi P, Comolombo P. Delivery of nasal powders of beta-cyclodextrin by insufflation. Pharmaceutical Research 1996;13(5):734–738.

19. Russo P, Sacchetti C, Pasquali I, Bettini R, Massimo G, Colombo P, Rossi A. Primary microparticles and agglomerates of morphine for nasal insufflation. Journal of Pharmaceutical Sciences 2006;95(12):2553–2561.

20. US FDA. Nasal Spray and Inhalation Solution, Suspension, and Spray Drug Products—Chemistry, Manufacturing, and Controls Documentation. Silver Spring, MD: US Food and Drug Administration; 2002. pp. 1–45.

3

Formulation of inhalation medicines

Daniela Traini and Paul M. Young

Respiratory Technology, The Woolcock Institute of Medical Research & The Discipline of Pharmacology, The University of Sydney, Sydney, Australia

3.1 Introduction

Chapter 2 highlighted the different approaches and commercial devices used to deliver medicines to the respiratory tract. For the routine treatment of nonsevere asthma, the two most commonly used devices are pressurized metered-dose inhalers (pMDIs) and dry powder inhalers (DPIs). As discussed in Chapter 2, a wide range of devices are available on the market and research within this area spans over 50 years. The technology underpinning the formulation of these systems is thus extensive. This chapter gives an insight into the principles and research governing the formulation of pMDI and DPI systems.

3.2 Pressurized metered-dose inhaler (pMDI) formulation

The formulation of active pharmaceutical ingredients in pressurized liquids is relatively complex, since the propellants used have low dielectric constants and are relatively inert. Historically, chlorofluorocarbons (CFCs) were used, but due to their ozone-depleting effect [1] they were phased out following the 1989 Montreal Protocol, and a replacement was required.

Hydrofluoroalkanes (HFAs) were identified as a potential alternative, since they were liquid propellants and were considered inert with respect to the environment [2]. Furthermore, HFAs had no potential toxicological issues [3].

Inhalation Drug Delivery: Techniques and Products, First Edition. Paolo Colombo, Daniela Traini, and Francesca Buttini.
© 2013 John Wiley & Sons, Ltd. Published 2013 by John Wiley & Sons, Ltd.

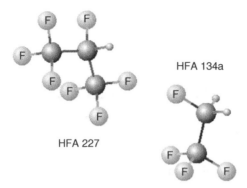

Figure 3.1 Chemical structures of HFA 227 and HFA 134a propellants.

Two HFAs are utilized in the formulation of pMDIs: namely, HFA 227 and HFA 134a. The nomenclature for their numbering (XYZ) is as follows: X = carbon atom number − 1; Y = hydrogen atom number + 1; Z = number of fluoride atoms. A letter following this sequence is sometimes used to indicate isomerism (as in HFA 134a), with increasing suffix value indicating decreased symmetry [4]. The structures of HFA 227 and HFA 134a are shown in Figure 3.1, and some of their basic physical properties are given in Table 3.1.

HFAs have low boiling points, high vapor pressures, and low dielectric constants. Although many of the physical properties of HFAs are similar to those of CFCs, direct translation of CFC formulations to HFA formulations was not possible. Historically, CFC formulations contained drug suspended in CFCs that were stabilized using surfactants. With the translation to HFA-based systems, it quickly became evident that the capacity for HFAs to solubilize these surfactants was not sufficient [5], and thus a stable flocculated system could not be formed. HFA pMDI systems thus had to be reformulated from scratch.

Table 3.1 Some physical properties of HFA 227 and HFA 134a [4]

	HFA 227	HFA 134a
Chemical name	1,1,1,2,3,3,3-heptafluoropropane	1,1,1,2-tetrafluoroethane
Chemical formula	CF_3-CFH-CF_3	CF_3CH_2F
MW	170.03	102.03
Boiling point (°C)	−16.5	−26.3
Vapor pressure (bar)	3.90	5.72
Dielectric constant	4.1	9.8
Liquid density (kg/m^3)	1408	1226
Solubility of water (g/kg 25 °C)	0.61	2.20
Solubility of ethanol (25 °C)	Miscible	Miscible

Formulations of HFA-based pMDI systems are generally categorized as either suspension or solution technologies. As the names suggest, suspension formulations contain a suspended drug formulation with or without a stabilizing agent, while solution-based formulations contain the drug solubilized in the HFA, with or without cosolvents, additives, and stabilizers. These approaches are discussed in more detail below.

3.2.1 Suspension technology

To highlight the importance of suspension stability in pMDI systems, one must consider the metering system. Unlike conventional pharmaceutical suspensions, pMDI systems have to be metered in such a way as to maintain the integrity of the bulk formulation while exposing a small amount of pressurized liquid to standard pressure. Upon exposure, this small volume rapidly expands and an aerosol is generated (Figure 3.2).

The important thing to consider here is the metering chamber. During storage, and prior to actuation, the bulk formulation has continual access to the metering chamber. Furthermore, the chamber volume is small; typically around 50 μL. Thus, in order to ensure a consistent dose, the formulation needs to either (1) remain as a homogeneous suspension or (2) be readily resuspended, such that a representative sample of the bulk can be introduced into the metering chamber.

Another important issue is the materials used. Components—including spring, metering chamber, canister wall, and gaskets—need to have a relatively low adhesion/absorption potential with respect to the drug. Furthermore, pressure-sealing gaskets need to have low water permeability and minimal swelling with respect to the liquids used in the formulation [6,7].

A review of any pharmaceutical textbook will provide a series of strategies for making a conventional suspension: utilizing the combination of Stokes' law and electro-double-layer forces to create a stable, flocculated system (i.e. utilizing the

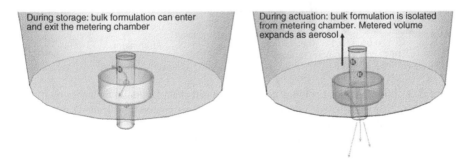

During storage: bulk formulation can enter and exit the metering chamber

During actuation: bulk formulation is isolated from metering chamber. Metered volume expands as aerosol

Figure 3.2 Schematic of a conventional pMDI valve before and after actuation

Derjaguin and Landau, Verwey and Overbeek (DLVO) theory). Unfortunately, HFAs have a low dielectric constant (Table 3.1) and thus will exert only a weak double-layer force on suspended particles.

To overcome such issues, formulations must either alter the fundamental properties of the drug and media (for example, by altering propellant/drug density or by altering the particle size, to reduce sedimentation velocity) or modify the contact dynamics between the particulate systems. While the former approach is enticing, changing the particle size or density is fraught with difficulties, since there is potential for Oswald ripening (i.e. crystal growth within the formulation) when the size of particles is reduced or cosolvents are added to a formulation.

Changing the contact dynamics between particles appears more straightforward, as surfactants are routinely used in pharmaceutical formulations to alter the interfacial tension, form colloid bridges, or act as a steric shell. While surfactants such as oleic acid were historically used in CFC formulations, they are not soluble in HFAs and require cosolvents, which may influence the stability and solubility of the suspended drug [5]. Luckily, potential replacements exist that are soluble in HFAs; these include long-chain polymers such as polyethyleneglycol (PEG) or polyvinylpirrolidone (PVP), fluorinated carboxylic acids/ester surfactants, and hydrophilic surfactants.

These HFA-soluble surfactants have the ability to generate stable suspensions via reduced particle–particle and particle–pMDI component interactions. Figure 3.3a shows the influence of PEG concentration and molecular weight on the adhesion between drug particles in a model propellant, for example [8].

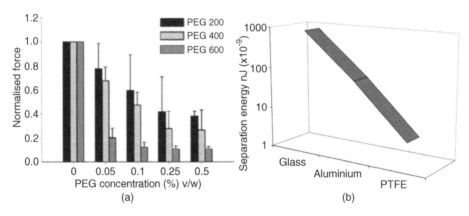

Figure 3.3 (a) Influence of polymer concentration and chain length on the adhesion of drug particles in an HFA-based system. Reproduced with permission from Traini D, Young PM, Rogueda P, Price R. Investigation into the influence of polymeric stabilizing excipients on inter-particulate forces in pressurised metered dose inhalers. Int J Pharm. 2006. (b) Adhesion of drug to different canister materials. Based on data from Traini D, Young PM, Rogueda P, Price R. Investigation into the influence of polymeric stabilizing excipients on inter-particulate forces in pressurised metered dose inhalers. Int J Pharm. 2006.

In addition to altering the "surface chemistry" of the drug particles via addition of surfactants, the surface energy of the pMDI canister components may be changed by material selection or coating technologies. This can be seen in Figure 3.3b when comparing the adhesion of drug to glass, aluminum, or polytetrafluoroethylene (PTFE). The careful selection of canister materials has already found use in commercial pMDI products and can most notably be observed in the Ventolin HFA product, which uses a fluorinated canister-coated pMDI with no additional surfactants [9].

3.2.2 Solution technology

Solution-based pMDIs do not encounter the same issues associated with suspensions (i.e. sedimentation, caking, particle adhesion, Oswald ripening, etc.); however, they have their own unique issues, which need to be considered. Solution pMDIs are effectively dispersions in which individual drug molecules interact with the HFA/formulation at the molecular level. Such systems will thus have a higher potential for chemical degradation when compared to their suspension counterparts. In addition, drug molecules conventionally used in pMDIs are not readily soluble in HFAs and thus require a cosolvent. Finally, when solubilized, these systems require good physical stability and must not precipitate when exposed to variations in temperature.

As outlined in Table 3.1, ethanol is miscible in HFAs and is also a good solvent for many hydrophobic pharmaceutical drugs. Thus ethanol may be used as a cosolvent.

An example of an ethanol cosolvent/HFA formulation is 3Ms beclomethosone dipropionate formulation: Qvar. Interestingly, Qvar was reported as having improved respiratory deposition (50–60% in human studies) when compared to previous suspension pMDIs [10]. This is most likely due to the nature of solution formulations compared with suspensions. In suspensions, the particle size of the inhaled drug will be dependent on the size of the micronized material. However, in solution-based pMDIs, the inhaled particle size is dependent on the concentration of drug in the propellant and the nature of the droplet formulation during aerosolization (i.e. small droplets result in smaller particles). By controlling parameters such as orifice diameter and vapor pressure, the final particle size distribution may thus be controlled, along with the respiratory deposition profile (Figure 3.4). For example, by reducing the orifice size (the hole through which the aerosol droplets form when they exit the pMDI), a reduction in the droplet diameter will be observed. This results in a decrease in the final particle size and an increase in respirator deposition.

In general, solution-based pMDIs, utilizing volatile cosolvents, result in higher fine-particle fractions, due to the small particle size of the dried aerosol. Indeed, products such as Qvar result in particle sizes in the range of 0.8–1.2 μm. Although suitable for respiratory delivery, this size of aerosol will be mainly deposited

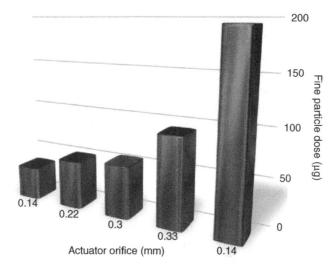

Figure 3.4 Influence of orifice diameter on the fine particle dose of beclomethasone diproprionate (BDP) from a solution pMDI. Reproduced with permission using data from Ganderton D, Lewis D, Davies R, Meakin B, Brambilla G, Church T. Modulite (R): a means of designing the aerosols generated by pressurized metered dose inhalers. Respiratory Medicine 2002;96:S3–S8

in the alveoli. For some formulations, this may not be suitable, either for clinically relevant reasons or when equivalence to other inhalation medicines is required.

The particle size of solution-based pMDIs may be altered via the addition of nonvolatile agents [11]. Nonvolatile agents are soluble in the HFA-cosolvent system; however, they will not evaporate during the aerosolization process. Since evaporation is dependent on the initial solution composition, the integration of nonvolatiles into the formulation will result in the final aerosol containing both drug and nonvolatile additive. Therefore, by increasing the nonvolatile concentration, one may increase the final particle size. Nonvolatile components include glycerol and polyethelene glycol, and their application can be seen in the Chiesi Modulite technology, where increasing drug concentration and nonvolatile concentration results in an increase in the mass median aerodynamic diameter (MMAD) (Figure 3.5).

3.3   Dry powder inhaler (DPI) formulation

The formulation of an efficient DPI is based on the integration of a device with a powder formulation. Two powder formulation routes are commonly found in the majority of DPIs: carrier-based systems and agglomerated systems. Although other formulation options exist (for example, DPI formulations may be prepared as

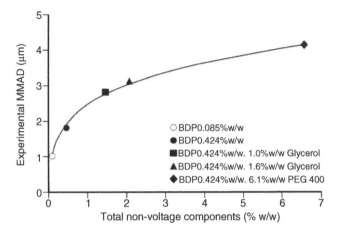

Figure 3.5 Effect of the total nonvolatile component on the MMAD of BDP pMDI solution formulations in HFA 134a. Reproduced with permission from Ganderton D, Lewis D, Davies R, Meakin B, Brambilla G, Church T. Modulite (R): a means of designing the aerosols generated by pressurized metered dose inhalers. Respiratory Medicine 2002;96:S3–S8

cospray dried composites), for the purposes of this chapter only the two main approaches and the factors that influence their performance are discussed.

3.3.1 Carrier technology

The reason for the use of carrier materials is the poor flow properties of the micron-sized drug, in combination with the related difficulty of metering such small volumes/masses. Compared to conventional solid dosage forms, DPIs generally have very small therapeutic doses (e.g. formoterol-fumarate-dihydrate $\leq 12\,\mu g$). By mixing the drug with a larger carrier such as lactose, it becomes easier to meter the drug and aerosolize the bulk powder during inhalation (Figure 3.6).

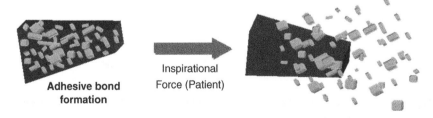

Figure 3.6 Schematic of a DPI carrier-based system

Carrier formulations are based on the principle of ordered mixing [12]. In comparison to random mixing, ordered mixing results in a "predictable" component distribution. Since the drug particles are much smaller than the carrier (usually one order of magnitude), the smaller drug particles preferentially adhere to the carrier, resulting in an adhesive mix. During inhalation, the energy imparted by the patient must overcome the adhesive bond formation, so that the drug particles can be liberated from the carrier to penetrate the respiratory tract. However, the force of adhesion between the drug and the carrier may be greater than the energy supplied, and thus the drug will remain on the carrier, to be swallowed, after impaction in the throat instead then inhaled. Such an achievement is not as simple as it first seems. To put this in perspective, the force of adhesion between a drug particle and carrier may be calculated from the empirical formula (Equation 3.1) describing the adhesion (F_{adh}) of a solid sphere (with diameter d) to a planar surface [13], and compared to the force encountered by gravity, centrifugal motion, or airflow (Figure 3.7).

$$\text{Empirical adhesion force}: F_{adh} = 0.063\,2d[1 + 0.009(\%RH)] \qquad (3.1)$$

From Figure 3.7, it can be seen than the force imparted on a micron-sized particle is generally less than the force of adhesion. Although this is a very simplistic

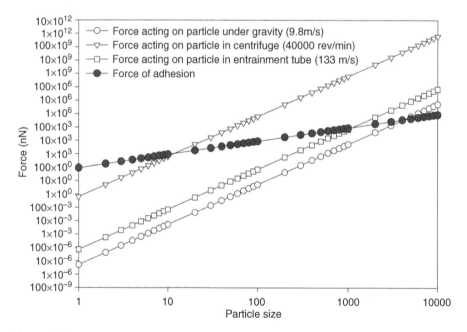

Figure 3.7 Theoretical force required to removal a hydrophobic sphere (of specific diameters) from a planar surface, plotted alongside theoretical forces acting on it due to gravity and linear accelerations

calculation, it does go some way to describing the poor efficiency of DPIs, where fine-particle fractions <20% are often observed [14].

In general, however, the ability to have a consistent dose and good powder flowability means carrier-based systems are often the preferred formulation option. The development of technologies to improve aerosol performance by modifying the carrier is thus a core research theme, and many inhalation and custom grades of lactose are available for DPI formulations [15–19].

It is believed that there are two factors which dominate drug–carrier adhesion and performance: (1) the presence of "active sites" and (2) drug/fine microagglomeration. Active sites may be considered as areas on the lactose surface that have a high adhesion potential; drug particles adhered to these areas will thus be more difficult to remove. Active sites may be caused by morphological features (i.e. pits and clefts on the surface), surface roughness, amorphous content, and/or variations in surface energy (due to polymorphism, surface chemistry, etc). Figure 3.8 (left pane) shows drug particles caught in active sites on the surface of a lactose carrier and the corresponding effect on performance, where drug particles are not released from the surface until all these sites are filled [20]. Furthermore, in this example it can be seen

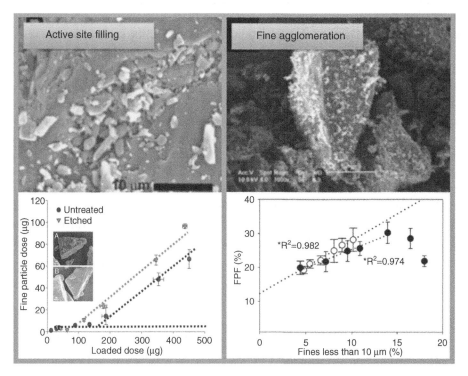

Figure 3.8 Active-site and fine-agglomeration theories. SEM images of the effect, with corresponding influence on aerosolization performance. Reproduced with permission from [21–23]

that etching the surface may reduce the number of active sites. Since there are a lower number of sites to fill, aerosolization performance increases at a lower threshold dose [21].

In comparison, Figure 3.8 (right pane) shows the influence of lactose fines on the performance. Fines may be considered as excipient particles with a similar size to that of the drug. Since lactose are a similar size, they form ordered mixes with the larger carrier, but simultaneously may form strong adhesion/cohesion bonds with other fines or drug particles [22–24]. In addition, this mechanism has been reported to be a dose-dependent effect, where an increase in fine concentration results in an increased likelihood of fine–fine/fine–drug contact and thus agglomeration [25]. The formation of these microagglomerates, multiplets, or drug–fine clusters results in an increase in the effective mass of the active drug and thus a decrease in surface area-to-mass ratio. Ultimately, the force imparted on the drug particles during the inhalation process becomes greater (Figure 3.7) and the entrained particle system may disperse to penetrate the lung. The outcome of the active-site and fine-agglomeration theories is the development of lactose, which is processed to reduce the number of active sites and/or the development of fine lactose grades and their addition into formulations.

As discussed, the force of adhesion between a drug particle and carrier is high due to the drug's high surface area-to-mass ratio. However, the magnitude of the adhesion force will be directly proportional to the sum of the surface free energies of each component and the effective contact area between the two contiguous surfaces. For example, the adhesion of a drug particle on a flat surface (Figure 3.9, image 2) will be greater than on a surface covered in many small asperities (Figure 3.9, image 3), since the overall contact area, and thus adhesion force, is reduced. Interestingly, when irregular surfaces with pits and clefts are observed, the adhesion force is likely to increase (as discussed in the active-site theory), due to increased contact area (Figure 3.9, image 1). Many studies have reported methods to alter the surface roughness of carriers; these include wet-milling [26], surface-etching [27], controlled crystallization [28], and granulation [29].

Modification of the surface roughness of a carrier may result in a change in contact area with the drug; however, this is unlikely to affect the intrinsic surface energy, which is dependent on the chemical species and the arrangement of

Adhesion and interactive force of lactose with drug

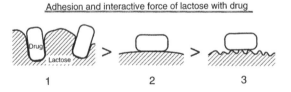

Figure 3.9 Influence of carrier roughness on drug adhesion force

molecules on the surface. While it is neither easy nor recommended to alter the integral chemistry of the active pharmaceutical ingredient, the surface chemistry of the carrier may be altered without risk of adverse therapeutic effect.

The surface energy may be changed via three main process: (1) alternative carriers may be used; (2) an existing carrier may be engineered as a different polymorph or different crystal habit; or (3) an existing carrier may be processed with a ternary agent to alter the surface chemistry (i.e. coating).

Studies by Tee et al. [30], Steckel & Bolzen [31], and Traini et al. [32] have investigated various sugars and sugar alcohols for DPI formulation (from polyols, such as mannitol and erythritol, to more complex disaccharides, such as sucrose and lactose). While many of these have shown potential, lactose remains the most common DPI excipient, most likely due to its regulatory acceptance [15].

The crystallization and manufacture of lactose in different polymorphic forms also offers an opportunity to alter surface chemistry and thus DPI performance. For example, work by Traini et al. [33] investigated the relationship between different crystal packing arrangements of lactose, their surface energy, and their performance as DPI carriers (Figure 3.10). In general, an increase in carrier surface energy resulted in an increased drug–carrier adhesion. The result of such an effect was an overall decrease in drug aerosolization performance as surface energy increased. The use of polymorphism to alter the performance of DPIs is limited, however, since the number and the form of the polymorph and crystal

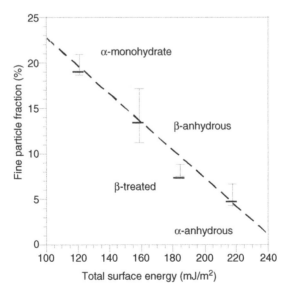

Figure 3.10 The influence of lactose polymorphism on surface energy and drug aerosolization efficiency

habits are finite (limited by the number of packing combinations and crystal growth conditions).

An alternative to this approach is to alter the surface chemistry by coating with a ternary compound. For example, magnesium stearate, commonly used as a lubricant in tablet manufacture, may be coated on to the surface of lactose via high-shear wet-mixing [26] or mechanofusion [34]. Since ternary agents are used to alter the adhesion force between drug and carrier, they are often referred to as force-control agents (FCA). Apart from magnesium stearate, force control agents studied include leucine, sucrose stearate, and sodium stearate. Chiesi's beclometasone Pulvinal product uses this approach by including magnesium stearate in the formulation [35].

3.3.2 Agglomerate technology

In comparison to carrier-based formulations, agglomerate-based formulations contain no large inert material, but are spheronized aggregates containing many micron-sized particles, of suitable size for inhalation. By making controlled aggregates, powder flow and dispersability are maintained, allowing easy metering and powder aerosolization. Where carrier systems require the liberation of drug from the carrier surface during the inhalation process, agglomerate systems rely on the inspirational energy being sufficient to break up the powder network (resulting in primary particles of respiratory size) (Figure 3.11).

Agglomerated systems have good flow and metering properties, since the agglomerate size is large compared to the primary micron-sized powder. Astra Zeneca's Turbuhaler formulations utilize this technology in, for example, Pulmicort budesonide formulations (containing 100–400 μg/dose) [36]. However, when the formulation dose becomes smaller than 100 μg, the agglomerate size becomes too small to maintain good flow properties. Low-dose formulations are thus agglomerated as binary systems containing both drug and micronized excipient.

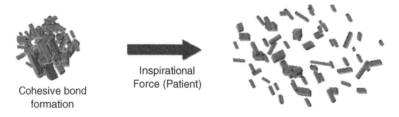

Cohesive bond formation

Inspirational Force (Patient)

Figure 3.11 Agglomerate formation and dispersion

3.4 Conclusion

Research and development into the formulation of both pMDI and DPI medicines is a continually evolving area. This chapter has focused on the core formulation strategies, to give an insight into the complexities of this subject area. It is clear that there is room for significant improvement in the current state of the art, in order to achieve improved lung deposition in future formulations.

References

1. Molina MJ, Rowland FS. Stratospheric sink for chlorofluoromethanes—chlorine atomic-catalysed destruction of ozone. Nature 1974;249(5460):810–812.
2. Jones DS, McCoy CP. Heptafluoropropane (HFC). In Rowe RC, Sheskey PJ, Weller PJ, editors. Handbook of Pharmaceutical Excipients. London: Pharmaceutical Press; 2009. pp. 303–305.
3. Alexander DJ, Libretto SE. An overview of the toxicology of Hfa-134a (1,1,1,2-tetrafluoroethane). Human & Experimental Toxicology 1995;14(9):715–720.
4. Solvay Fluor. Solkane® 227 Pharma and 134a Pharma Data Sheet. http://www.solvaychemicals.com/EN/products/Fluor/Hydrofluorocarbons_HFC/Solkane227pharma.aspx and http://www.solvaychemicals.com/EN/products/Fluor/Hydrofluorocarbons_HFC/Solkane134apharma.aspx. 2010 [last accessed Jan 17, 2010].
5. Smyth HDC. The influence of formulation variables on the performance of alternative propellant-driven metered dose inhalers. Advanced Drug Delivery Reviews 2003;55(7):807–828.
6. Atkins PJ, editor. The Development of New Solution Metered Dose Inhaler Delivery Systems. Buffalo Grove, IL: Proceedings of Respiratory Drug Delivery; 1990.
7. Kontny MJ, Destefano G, Jager PD, Mcnamara DP, Turi JS, Vancampen L. Issues surrounding MDI formulation development with non-CFC propellants. Journal of Aerosol Medicine 1991;4(3):181–187.
8. Traini D, Young PM, Rogueda P, Price R. Investigation into the influence of polymeric stabilizing excipients on inter-particulate forces in pressurised metered dose inhalers. International Journal of Pharmaceutics 2006;320(1–2):58–63.
9. Ashurst ICW (GB), Herman CSR (NC), Li L SP (NJ), Riebe MTR (NC), inventors; Glaxo Wellcome Inc. (Research Triangle Park, NC); Glaxo Group Limited (Greenford, GB), assignee. Metered dose inhaler for salmeterol. US patent 6143277. 1996.
10. Leach CL. Targeting inhaled steroids. International Journal of Clinical Practice 1998;96:23–27.
11. Ganderton D, Lewis D, Davies R, Meakin B, Brambilla G, Church T. Modulite (R): a means of designing the aerosols generated by pressurized metered dose inhalers. Respiratory Medicine 2002;96:S3–S8.

12. Hersey JA. Ordered mixing—new concept in powder mixing practice 17. Powder Technology 1975;11(1):41–44.

13. Hinds WC. Aerosol Technology. New York, NY: John Willey & Sons, Ltd; 1999.

14. Smith IJ, Parry-Billings M. The inhalers of the future? A review of dry powder devices on the market today. Pulmonary Pharmacology & Therapeutics 2003;16 (2):79–95.

15. Edge S, Kaerger S, Shur J. Lactose, inhalation. In Rowe RC, Sheskey PJ, Weller PJ, editors. Handbook of Pharmaceutical Excipients. London: Pharmaceutical Press; 2009. pp. 362–4.

16. Edge S, Kaerger S, Shur J. Lactose, anhydrous. In Rowe RC, Sheskey PJ, Weller PJ, editors. Handbook of Pharmaceutical Excipients. London: Pharmaceutical Press; 2009. pp. 359–361.

17. Edge S, Kaerger S, Shur J. Lactose, monohydrate. In Rowe RC, Sheskey PJ, Weller PJ, editors. Handbook of Pharmaceutical Excipients. London: Pharmaceutical Press; 2009. pp. 364–369.

18. Edge S, Kaerger S, Shur J. Lactose, spray-dried. In Rowe RC, Sheskey PJ, Weller PJ, editors. Handbook of Pharmaceutical Excipients. London: Pharmaceutical Press; 2009. pp. 376–378.

19. Kaerger SJ, Price R, Young PM, Tobyn MJ. Carriers for DPIs: formulation and regulatory challenges. Pharmaceutical Technology Europe 2006; October edition:25–30.

20. Young PM, Edge S, Traini D, Jones MD, Price R, El-Sabawi D, et al. The influence of dose on the performance of dry powder inhalation systems 1. International Journal of Pharmaceutics 2005;296(1–2):26–33.

21. El-Sabawi D, Edge S, Price R, Young P. Continued investigation into the influence of loaded dose on the performance of dry powder inhalers: surface smoothing effects. Drug Development and Industrial Pharmacy 2006;32(10):1135–1138.

22. Islam N, Stewart P, Larson I, Hartley P. Lactose surface modification by decantation: are drug–fine lactose ratios the key to better dispersion of salmeterol xinafoate from lactose-interactive mixtures? 2. Pharmaceutical Research 2004;21(3): 492–499.

23. Jones MD, Price R. The influence of fine excipient particles on the performance of carrier-based dry powder inhalation formulations. Pharmaceutical Research 2006;23(8):1665–1674.

24. Lucas P, Anderson K, Staniforth JN. Protein deposition from dry powder inhalers: fine particle multiplets as performance modifiers. Pharmaceutical Research 1998;15(4):562–569.

25. Young PM, Traini D, Chan HK, Chiou H, Edge S, Tee T. The influence of mechanical processing dry powder inhaler carriers on drug aerosolisation performance. Journal of Pharmaceutical Sciences 2007;96(5):1331–1341.

26. Ferrari F, Cocconi D, Bettini R, Giordano F, Sant P, Tobyn MJ et al. The surface roughness of lactose particles can be modulated by wet-smoothing using a high-shear mixer. AAPS PharmSciTech 2004;5(4):1–6.

27. El-Sabawi D, Price R, Edge S, Young PM. Novel temperature controlled surface dissolution of excipient particles for carrier based dry powder inhaler formulations 1. Drug Development and Industrial Pharmacy 2006;32(2):243–251.

28. Zeng XM, Martin GP, Marriott C, Pritchard J. The use of lactose recrystallised from carbopol gels as a carrier for aerosolised salbutamol sulphate. European Journal of Pharmaceutics and Biopharmaceutics 2001;51(1):55–62.

29. Young PM, Roberts D, Chiou H, Rae W, Chan HK, Traini D. Composite carriers improve the aerosolisation efficiency of drugs for respiratory delivery. Journal of Aerosol Science 2008;39(1):82–93.

30. Tee SK, Marriott C, Zeng XM, Martin GP. The use of different sugars as fine and coarse carriers for aerosolised salbutamol sulphate. International Journal of Pharmaceutics 2000;208(1–2):111–123.

31. Steckel H, Bolzen N. Alternative sugars as potential carriers for dry powder inhalations. International Journal of Pharmaceutics 2004;270(1–2):297–306.

32. Traini D, Young PM, Jones MD, Edge S, Price R. Comparative study of erythritol and lactose monohydrate as carriers for inhalation: atomic force microscopy and in vitro correlation. European Journal of Pharmaceutical Sciences 2006;27(2–3): 243–251.

33. Traini D, Young PM, Thielmann F, Acharya M. The influence of lactose pseudo-polymorphic form on salbutamol sulfate-lactose interactions in DPI formulations. Drug Development and Industrial Pharmacy 2008;34(9):992–1001.

34. Kumon M, Suzuki M, Kusai A, Yonemochi E, Terada K. Novel approach to DPI carrier lactose with mechanofusion process with additives and evaluation by IGC. Chemical & Pharmaceutical Bulletin 2006;54(11):1508–1514.

35. Chiesi. Beclometasone Pulvinal Product Data Sheet. http://www.medicines.org.uk/ EMC/medicine/21030/SPC/Pulvinal%20Beclometasone%20Inhaler%20100,200% 20and%20400%20micrograms/.

36. Wetterlin K. Turbuhaler—a new powder inhaler for administration of drugs to the airways. Pharmaceutical Research 1988;5(8):506–508.

4

Novel particle production technologies for inhalation products

Hak-Kim Chan[1] and Philip Chi Lip Kwok[2]

[1]Advanced Drug Delivery Group, Faculty of Pharmacy, The University of Sydney, Sydney, Australia
[2]Department of Pharmacology and Pharmacy, LKS Faculty of Medicine, The University of Hong Kong, Hong Kong, China

4.1 Introduction

Manufacturing of suspension-type metered-dose inhaler (MDI) and dry powder inhaler (DPI) products for pulmonary drug administration requires powders with desirable characteristics, including particle size, shape, density, crystallinity, surface morphology, and surface energy. These characteristics will affect the physical stability of the aerosol formulations and the interactions between different particles and between particles and the inhaler during aerosolization. Drug powders for inhalation have traditionally been produced by crystallization followed by milling to reduce the particles to the size range required for aerosol formulation and delivery to the lungs. While such a production approach to inhalation products may have been sufficient in the past, it is not suitable for producing powders with the required flow and dispersion characteristics to meet the need for enhanced powder performance. Various methods that have been explored or are in an advanced stage of development are examined in this chapter.

4.2 Conventional crystallization and milling

Crystallization involves several main steps: supersaturation, nucleation, and crystal growth. Both nucleation and crystal growth depend on the level of supersaturation.

Inhalation Drug Delivery: Techniques and Products, First Edition. Paolo Colombo, Daniela Traini, and Francesca Buttini.
© 2013 John Wiley & Sons, Ltd. Published 2013 by John Wiley & Sons, Ltd.

Industrial crystallization usually takes place within chemical reactors, which can be scaled up for large-batch production. Since it is not easy to achieve a uniform supersaturation level inside the reactor, conventional crystallization suffers from disparity in the rate of nucleation and the subsequent crystal growth. This will lead to poor control over the particle shape and size distribution. In general, this process tends to produce crystals >10 μm, with a broad size distribution, which will not be suitable for inhalation to the lungs.

After crystallization, the precipitate is collected by filtration and is dried in an oven at a set temperature. To reduce the particle size of the crystals, milling (e.g. fluid-energy/air-jet-milling or ball milling) is commonly employed in the pharmaceutical industry. In ball milling, powder contained inside a drum is grounded by rolling balls of various sizes as the drum rotates. In fluid-energy milling, compressed air is fed along with the powder into the mill, where the air accelerates and causes attrition of the particles by collision. Due to the high energy imparted to the particles in the milling process, they become partially amorphous, with reduced crystallinity (and hence attain a higher energy state than the fully crystalline materials). These particles can also carry electrostatic charge, resulting from friction or contact with the mill surface—a process known as trioboelectrification. Having higher surface energy and charge, these particles are very cohesive, making the powder difficult to handle. Furthermore, the powder will recrystallize from the partially amorphous state under elevated temperature and humidity conditions, causing potential instability problems during manufacturing and storage. An additional problem is that malleable materials such as inhaled steroid drugs (e.g. triamcinolone acetonide) are difficult to mill [1]. Despite these limitations, crystallization and milling are established techniques and have been widely used in the past for inhalation products [2–4].

However, because of the potential limitations of the conventional production process, alternative methods have emerged for the manufacture of powders with enhanced performance, increasing the amount of fine particles in the aerosol—with less device retention—and reducing dose variability.

4.3   Specialized milling

Micronization may produce partially amorphous materials with a surface charge, causing particle agglomeration. These problems can be dealt with by specialized milling methods.

4.3.1   Fluid-energy milling at elevated humidity

In order to reduce the amorphous content in the material produced by milling, the milling can be carried out at elevated humidity. The absorbed moisture functions as a plasticizer to lower the glass transition temperature of the amorphous material, hence facilitating in situ crystallization during the milling process. The milled products have been reported to be predominantly crystalline, with particle size distributions similar to those produced by the conventional milling process [5]. The setup involves controlling the relative humidity (e.g. 30–70%) of the milling chamber by humidifying the feed gas (e.g. by superheated steam, to minimize condensation) used to mill the powder [5].

4.3.2 Wet-milling nanotechnology

Nanocrystal Technology (Elan Drug Technologies, King of Prussia, PA, USA) is an aqueous-based milling process used to reduce particle size to below 400 nm. A conventional ball mill can be used for the process, and the materials selected for the grinding media (e.g. glass, zirconium oxide) have been reported not to be crucial [6]. However, the grinding media should preferably be ≤1 mm in order to be effective in attrition and to impart less wear to the mill [6]. Milling in a water medium causes amorphous-to-crystalline transformation, and the resulting powder is physically more stable than that produced by dry milling.

Budesonide and other compounds were milled to nanosized particles, which were then spray- or freeze-dried for use in MDI, DPI, or nebulizer systems [7,8]. A surface modifier (e.g. PVP, lecithin, cellulose derivatives) is usually added during or after milling to prevent agglomeration of the nanoparticles. Although these stabilizers can be of generally-recognized-as-safe (GRAS) materials, long-term inhalation can be a concern unless proven safe. Another major drawback of wet milling is that, depending on the type of mill and the drug, a lengthy processing time may be required (5 days or longer).

Another process used for wet milling is high-pressure (piston-gap) homogenization, which uses cavitation forces and impact or shear forces to reduce particle size. In this technique, the drug powder to be milled is dispersed in an aqueous solution containing a surface modifier (surfactant or polymer) and then stirred at high speed to form a suspension. It is then milled in a piston-gap homogenizer to <5 μm. This method has been used to produce budesonide nanosuspension containing drug particles of 500–600 nm suitable for nebulization [9]. In addition to the requirement for preprocessing the powder into a suspension, low suspension viscosity (hence low solid loading of <10%) may be a limitation of this method.

4.4 Solvent precipitation

Inhalable particles can potentially be obtained by rapid precipitation from aqueous solutions using antisolvents. A number of processes have emerged to control the nucleation rate and crystal growth, so as to reproducibly generate particle size in the micron range for aerosol delivery.

4.4.1   Sono-crystallization

Ultrasonic radiation has been applied to control the precipitation process [10]. The setup can simply comprise an ultrasound probe in a mechanically stirred reaction tank, where the antisolvent is mixed with the drug solution to precipitate the fine drug particles. The ultrasound frequency is crucial and 20–25 kHz (or higher) has been reported to be suitable for setups similar to the one shown in Figure 4.1 [10]. The ultrasonic waves help to nucleate and crystallize particles more evenly and faster. Nucleation and crystal-growth processes can be achieved in a short time because ultrasound reduces the width of the metastable zone and so nucleation can start at a lower level of supersaturation.

The sonication process is based on cavitation resulting from the constant creation, growth, and implosive disintegration of bubbles in liquid. Apart from the fast procedure, sonication offers further advantages: (1) it produces a smaller crystal size and narrower size distribution of the products than a conventional crystalliza-tion process, because crystal growth occurs at a lower supersaturation level, where initial growth is less rapid; (2) it is relatively cheap; (3) it can be carried out at ambient conditions; and (4) the reaction vessel required is relatively small and simply shaped, which makes the cleaning process easy.

The frequency of the ultrasonic wave affects the amount of cavitation (i.e. creation of bubbles) per unit volume per unit time and bubble size. The bubble

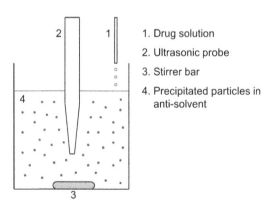

1. Drug solution
2. Ultrasonic probe
3. Stirrer bar
4. Precipitated particles in anti-solvent

Figure 4.1 Schematic diagram of the sono-crystallization process. Reproduced with permission from Chan H-K, Kwok PCL. Production methods for nanodrug particles using the bottom-up approach. Advanced Drug Delivery Reviews 2011;63:406–416.

Figure 4.2 Inhalable sodium chloride particles produced by sono-crystallization. Adapted with permission from Tang P, Chan H-K, Tam E, de Gruyter N, Chan J. Preparation of NaCl powder suitable for inhalation. Industrial & Engineering Chemistry Research 2006;45 (12):4188–4192. Copyright (2006) American Chemical Society

radius is inversely proportional to the frequency. At a given power, the amount of cavitation increases with increasing frequency, but the bubbles become smaller. Their implosions result in less energy being released, because bubble energy is proportional to the square of the bubble radius. Input power (amplitude) determines the size of the bubbles created. At a constant frequency, higher power (amplitude) will create bigger bubbles. When these bubbles burst, the intensity of the cavitation energy released is higher. At higher frequency, although the cavitation energy intensity is low, the number of cavitation implosions per unit volume per unit time is high. This could possibly lead to the enhancement of localized micromixing and hence to a narrow distribution of smaller particle size, as observed in sodium chloride particles for inhalation [11] (Figure 4.2).

Examples of antiasthmatic drugs prepared using sono-crystallization techniques include fluticasone propionate and salmeterol xinafoate [10]. This method should be applicable to the preparation of other inhalable compounds, such as salbutamol sulfate, beclomethasone dipropionate, budesonide, and formoterol fumarate.

In other applications, ultrasound has been employed either to induce crystallization inside dispersed solution droplets in a liquid medium, or to produce a stable emulsion followed by crystallization in the droplets induced by mass or heat transfer (such as cooling or diffusion of an antisolvent into the droplets) [12].

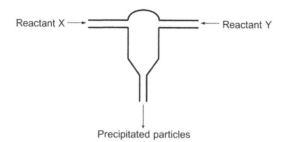

Figure 4.3 Schematic diagram of the opposing impinging liquid-jet process. Reproduced with permission from Chan H-K, Kwok PCL. Production methods for nanodrug particles using the bottom-up approach. Advanced Drug Delivery Reviews 2011;63:406–416.

4.4.2 Microprecipitation by opposing liquid jets and tangential liquid jets

In this method, precipitation occurs in a region of extreme turbulence and intense mixing, created by a jet of reactant liquid (e.g. drug solution) and a jet of another reactant liquid (e.g. antisolvent) coming through two facing nozzles mounted in a small chamber (Figure 4.3) [1]. As the two liquid jets mix, they react and cause the drug to precipitate as fine particles. The crucial process parameters include the speed of the liquid jets and the concentration of the reactants. A high jetstream speed or a high drug concentration have been found to give finer particles but a higher residual solvent level, while a low speed or low concentration produce the opposite effect [1]. The volume ratio of the reactants is also expected to affect the precipitation process.

A requirement for the operation of the opposing-liquid-jets process is that the jets must have equal velocity, otherwise one jet will be pushed back by the other. This can severely limit the flexibility of the process. To get around this problem, the process has been modified to form a four-jet swirling configuration for enhanced mixing (Figure 4.4).

4.4.3 High-gravity controlled precipitation

Uniform mixing will lead to uniform nucleation and crystal growth. There are two characteristic time parameters in crystallization: the induction time (τ) and the micro-mixing time (t_m). The induction time establishes a steady-state nucleation rate (normally in μs to ms). When $t_m \ll \tau$, the nucleation rate will be nearly uniform spatially, and the particle size distribution can be controlled at a uniform level. This has been achieved using high-gravity technology, which utilizes a rotating packed bed to intensify mass and heat transfer in multiphase systems (Figure 4.5) [13].

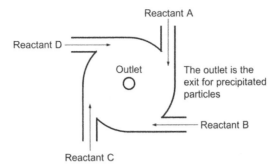

Figure 4.4 Schematic diagram of the four-jet swirling process. Reproduced with permission from Chan H-K, Kwok PCL. Production methods for nanodrug particles using the bottom-up approach. Advanced Drug Delivery Reviews 2011;63:406–416.

During rotation, the fluids going through the packed bed are spread and split into thin films, threads, and very fine droplets under the high shear created by the high gravity. This results in intense micro-mixing between the fluid elements of one to three orders of magnitude.

The process can be operated in two modes: reactive or antisolvent. Inhaled drugs such as salbutamol sulfate have been produced using both methods. In the reactive mode, the reactants (e.g. salbutamol base and sulfuric acid) entering the equipment will react and form salbutamol sulfate, whereas in the antisolvent mode the reactants will

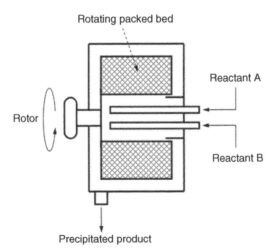

Figure 4.5 Schematic diagram of the high-gravity controlled process. Reproduced with permission from Chen J-F, Zhou M-Y, Shao L, Wang Y-Y, Yun J, Chew NYK, et al. Feasibility of preparing nanodrugs by high-gravity reactive precipitation. International Journal of Pharmaceutics 2004;269:267–274.

be salbutamol sulfate dissolved in water and an antisolvent (e.g. alcohol). On mixing, the latter will cause precipitation of the drug out of its aqueous solution [14,15].

4.5   Spray-drying and related droplet evaporation methods

4.5.1   Spray-drying

Spray-drying was explored in the 1980s as an alternative means of making fine particles with desirable flow and dispersion characteristics, without the need to use coarse carriers or form soft pellets. Antiasthmatic drugs, including salbutamol sulfate, terbutaline sulfate, isoprenaline sulfate, and sodium cromoglycate, were investigated [16–18]. However, it was not until the early 1990s when the potential of the pulmonary route for therapeutic protein delivery was recognized. An enormous effort was then focused on the spray-drying of pharmaceuticals.

In spray-drying, a drug solution is atomized to fine droplets, which are evaporated in a warm air current to form dry particles (Figure 4.6), which can be solid or partially hollow (Figure 4.7) [19,20]. Although the drying air temperature can be relatively high (>100 °C), the actual temperature of the evaporating droplets is

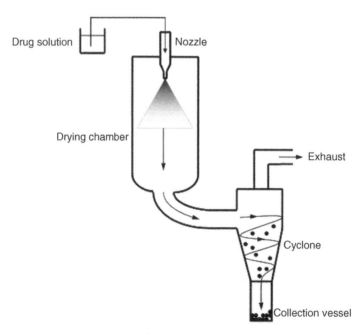

Figure 4.6   Schematic diagram of the spray-drying process

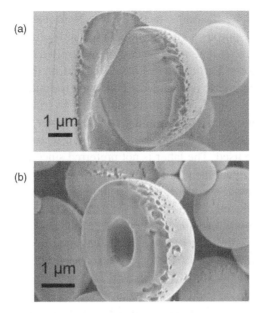

Figure 4.7 Spray-dried particles of bovine serum albumin, showing a solid (a) and a hollow. Reproduced with permission from Pharmaceutical research by AMERICAN ASSOCIATION OF PHARMACEUTICAL SCIENTISTS Reproduced with permission of SPRINGER NEW YORK LLC in the format use in a book/textbook via Copyright Clearance Center. (b) interior. Reproduced with permission from Pharmaceutical research by AMERICAN ASSOCIATION OF PHARMACEUTICAL SCIENTISTS Reproduced with permission of SPRINGER NEW YORK LLC in the format use in a book/textbook via Copyright Clearance Center.

significantly lower due to cooling by the latent heat of vaporization. Thus, thermal degradation of the active ingredient is not so much a concern as it first appears.

Instead of a single drug, when two active ingredients (e.g. a bronchodilator and a corticosteroid) are dissolved in a solution and spray dried, each of the resulting particles will contain both drugs, which allows the opportunity for combination drug therapy [21]. Furthermore, a drug and an excipient can be co-spray-dried to enhance the particle properties. For example, di-leucine and tri-leucine have been used to co-spray-dry with therapeutic peptides in order to improve aerosol performance by enriching the particle surface with these hydrophobic amino acids [22], while sodium cromoglycate has been co-spray-dried with hydrophobic amino acids, in particular leucine, to increase powder dispersibility [23].

In addition to drug production, spray-drying has been used to produce carrier particles; for example, spherical lactose carrier particles for budesonide formulation have been prepared by this method [24].

Spray-drying is not limited to aqueous solutions. Spray-drying of ethanolic solutions containing salbutamol sulfate and ipratropium bromide has been reported [25]. Pure beclomethasone dipropionate particles and beclomethasone dipropionate–hydroxypropylcellulose particles for controlled release in the lungs have been produced by the same method [26]. Nonaqueous systems have also been used to prepare porous particles suitable for aerosol delivery [27–29].

The properties of the spray-dried powders are controlled by both the process and formulation parameters. Earlier studies have looked into the effects of the active ingredient, atomizing nozzle type, powder collection technique, and droplet drying time [17,20,30]. The liquid feed can be atomized by rotary nozzles, two-fluid nozzles, or ultrasonic nozzles, depending on the droplet size required. Powder collection is usually achieved using a cyclone, but an electrostatic precipitator or a filter bag can also be used. The latter is not preferred since particulate of the filter material may contaminate the powder, and due to the airflow through the bag, the powder becomes compacted over time. Cyclones collect powders in a looser form, but the collection efficiency drops rapidly for particles below 1 or 2 μm, for which the production yield will become too low to be commercially viable. The driving force for drying is controlled by the water content and the difference between the inlet and outlet temperatures of the drying air. The drying time of the droplets depends on their residence time in the spray drier, which, in turn, is determined by the spray-drier dimensions and the drying airflow rate. It is important to note that these parameters are closely interrelated. Changing one process parameter will therefore lead to a change in the others. For example, while reducing the airflow will lengthen the time it takes for the droplets to evaporate, the drying efficiency will be reduced simultaneously, because less air will be available to evaporate the droplets. A lower drying airflow will also decrease the collection efficiency of the cyclone. However, higher airflow will evaporate the droplets more rapidly, resulting in a less-crystalline product due to there being insufficient time for crystallization. Thus, the usefulness of spray-drying in the preparation of stable fine particles is hampered. Spray-dried powders tend to be amorphous, as found with sodium cromoglycate and salbutamol sulfate [16,18]. This is a significant drawback for hydrophilic compounds as the amorphous materials will be physically unstable and will recrystallize at elevated temperatures and humidities. On the other hand, since hydrophobic compounds are not hygroscopic, they may not suffer the stability issue even at a high humidity if the glass transition temperature can be maintained well above the ambient temperature.

Amorphous content in spray-dried powders can be minimized by facilitating crystallization during particle formation, for example by lowering the drying rate to allow ordering of the molecules into the crystal lattice. This can be achieved by prolonging the drying time by inserting a secondary drying apparatus between the primary drying chamber and the cyclone [31]. A potential limitation of spray-drying is its unsuitability for substances sensitive to mechanical shear of atomization [30]. Drugs that are unstable to liquid–air interfaces and are decomposed by oxidation should be avoided, but the problem can be minimized by using an inert gas instead of air in the process, or, if feasible, using an antioxidant. It should be noted that, depending on the particle size range, powder nature, and cyclone collection efficiency, the process yield for inhalable particles can be unacceptably low. Despite these limitations, spray-drying has the distinct advantage of being a continuous process, incorporating both particle formation and drying into a single operation.

4.5.2   Controlled evaporation of droplets

Like spray-drying, controlled evaporation is a single-step continuous process involving the atomization of the drug solution into a carrier gas for drying [32,33]. Compared to spray-drying, this method provides better control over the temperature history and residence time of droplets. In the actual setup, the solution is atomized using an ultrasonic nebulizer. The droplets suspended in a carrier gas are then fed into a turbular flow reactor housed in a constant-temperature oven for evaporation. Since the feed rate and temperature are adjustable, the temperature history and residence time of the droplets can be controlled. The method has the potential to control the particle morphology and polymorphic form and has been used to produce beclomethasone dipropionate particles [34].

4.5.3   Evaporation of low-boiling-point solutions

This involves simply dissolving the active ingredient in a low-boiling-point organic solvent, then atomizing the solution and evaporating the resulting droplets to produce the dry particles [35]. The concept behind this approach is similar to both spray-drying and rapid expansion of supercritical fluid (SCF).

4.5.4   Spray freeze-drying

This was explored for pharmaceutical application in the early 1990s [36]. It involves spraying the drug solution into a freezing medium (usually liquid nitrogen), which turns the spray into frozen solution droplets, then lyophilizing to remove the ice via sublimation, leaving a powder behind (Figure 4.8).

Compared with spray-drying, this process produces light and porous particles with enhanced aerosol performance, and the production yield is almost 100%. It has been applied to the preparation of proteins such as rhDNase and anti-IgE antibody particles for inhalation [37,38]. However, it is an expensive process and is only justifiable for expensive drugs, as it requires the use of liquid nitrogen and the freeze-drying step is time-consuming. Furthermore, since the drug molecules are subject to shear stress in

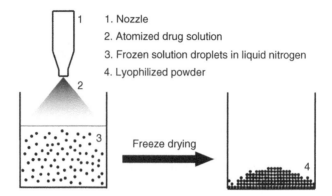

1. Nozzle
2. Atomized drug solution
3. Frozen solution droplets in liquid nitrogen
4. Lyophilized powder

Freeze drying

Figure 4.8 Schematic diagram of the spray freeze-drying process

the atomization step and cryogenic stress in the freeze-drying step, the process may not be suitable for molecules which are labile to these stresses.

4.6   Supercritical fluid (SCF) technology

It is well known that a compound can exist in three phases: solid, liquid, and gas. However, every compound also possesses a critical temperature and pressure above which no liquid phase formation can be formed by further compression. Under this condition, the gas and liquid have the same density and exist as a single phase. SCFs possess solvent density and solvent power intermediate to those of liquids and gases. These solvent properties have been explored to produce fine drug powders. Pharmaceutical SCF applications have mostly involved the use of super-critical carbon dioxide, due to its relatively low critical pressure (72.9 bar) and temperature ($31.1\,^\circ$C).

SCF technology has been applied in many different processes, but only the most fundamental ones will be included here. The first process is called rapid expansion of supercritical solution (RESS), in which an SCF is used as a solvent for the drug (i.e. solute) of interest (Figure 4.9a). The solution of drug in SCF is subject to rapid expansion as it passes through a nozzle at sonic speeds. During expansion, the density and solubilizing power of the SCF decrease dramatically, resulting in a high degree of solute supersaturation and subsequent precipitation of drug particles. For polar drugs which are poorly soluble in the relatively nonpolar supercritical carbon dioxide, this process is not suitable unless a cosolvent can be used. The second process involves the use of an SCF as an antisolvent to precipitate the drug from its organic solution (Figure 4.9b). This can be achieved by either spraying the drug solution as fine droplets into a flowing SCF stream or introducing the SCF gradually into the drug solution until precipitation occurs. SCF technology has successfully been used to prepare antiasthmatic compounds [39,40] and protein particles for inhalation [41–43].

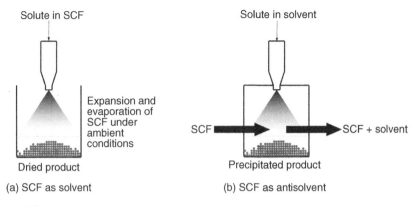

Figure 4.9 Schematic diagram of the supercritical fluid (SCF) process, using an SCF as (a) a solvent. Reproduced with permission from Chan H-K, Kwok PCL. Production methods for nanodrug particles using the bottom-up approach. Advanced Drug Delivery Reviews 2011;63:406–416. or (b) an antisolvent. Reproduced with permission from Chan H-K, Kwok PCL. Production methods for nanodrug particles using the bottom-up approach. Advanced Drug Delivery Reviews 2011;63:406-416.

4.7 Conclusion

A number of methods of powder production for the successful delivery of dry powder aerosols to the lungs have been reviewed, with a focus on the emerging technologies which can be used to overcome the limitations of current methods. Spray-drying has already been successfully employed for commercial-scale production for inhaler products. It is anticipated that other techniques, especially those that can easily be scaled up, may also become commercialized in the future.

Acknowledgments

This work is partly supported by research grants from the Australian Research Council.

References

1. Begon D, Guillaume P, Kohl M. Process for producing fine medicinal substances. WO patent 0114036. 2001.
2. Chiesi P. Conversion of an antiinflammatory steroid into a form administrable with an aerosol. DE patent 3018550. 1980.

3. Banholzer R, Sieger P, Kulinna C, Trunk M, Graulich MLA. Crystalline triotropium bromide monohydrate, method for producing the usage as inhalant powder. WO patent 0230928. 2002.

4. Clark AR, Hsu CC, Walsh AJ. Preparation of sodium chloride aerosol formulations. WO patent 9631221. 1996.

5. Vemuri NM, Brown AB, Authelin J-R, Hosek P. Milling process for the production of finely milled medicinal substances. WO patent 0032165. 2003.

6. Liversidge GG, Cundy KC, Bishop JF, Czekai DA. Surface modified drug nanoparticles. US patent 5145684. 1992.

7. Bosch HW, Ostrander KD, Cooper ER. Aerosols comprising nanoparticle drugs. US patent 2002/012294. 2002.

8. Ostrander KD, Hovey D, Knapp D, Parry-Billings M. Potential delivery advantages of spray-dried Nanocrystals™ colloidal budesonide with the Clickhaler®. In Dalby RN, Byron PR, Peart J, Farr SJ, editors. Respiratory Drug Delivery VII. Raleigh, NC: Serentec Press; 2000. pp. 447–449.

9. Jacobs C, Müller RH. Production and characterization of a budesonide nanosuspension for pulmonary administration. Pharmaceutical Research 2002;19 (2):189–194.

10. Lancaster RW, Singh H, Theophilus AL. Apparatus and process for preparing crystalline particles. WO patent 0038811. 2000.

11. Tang P, Chan H-K, Tam E, de Gruyter N, Chan J. Preparation of NaCl powder suitable for inhalation. Industrial & Engineering Chemistry Research 2006;45(12): 4188–4192.

12. Kaerger JS, Robert P. Processing of spherical crystalline particles via a novel solution atomization and crystallization by sonication (SAXS) technique. Pharmaceutical Research 2004;21:372–381.

13. Chen J-F, Zhou M-Y, Shao L, Wang Y-Y, Yun J, Chew NYK, et al. Feasibility of preparing nanodrugs by high-gravity reactive precipitation. International Journal of Pharmaceutics 2004;269:267–274.

14. Chiou H, Hu TT, Chan H-K, Chen J-F, Yun J. Production of salbutamol sulfate for inhalation by high-gravity controlled antisolvent precipitation. International Journal of Pharmaceutics 2007;331:93–98.

15. Hu TT, Chiou H, Chan H-K, Chen J, Yun J. Preparation of inhalable salbutamol sulphate using reactive high gravity controlled precipitation. Journal of Pharmaceutical Sciences 2008;97:944–949.

16. Vidgren MT, Vidgren, PA, Paronen TP. Comparison of physical and inhalation properties of spray-dried and mechanically micronized disodium cromoglycate. International Journal of Pharmaceutics 1987;35:139–144.

17. Forrester RB, Boardman TD. Inhalation pharmaceuticals. US patent 4590206. 1986.

18. Chawla A, Taylor KMG, Newton JM, Johnson MCR. Production of spray dried salbutamol sulfate for use in dry powder aerosol formulation. International Journal of Pharmaceutics 1994;108:233–240.

19. Heng D, Tang P, Cairney J, Chan H-K, Cutler D, Salama R, et al. Focused-ion-beam milling: a novel approach to probing the interior of particles used for inhalation aerosols. Pharmaceutical Research 2007;24:1608–1617.

20. Masters K. Spray Drying Handbook. New York, NY: Halsted Press; 1979.

21. Tajber L, Corrigan DO, Corrigan OI, Healy AM. Spray drying of budesonide, formoterol fumarate and their composites. I. Physicochemical characterisation. International Journal of Pharmaceutics 2009;367:79–85.

22. Lechuga-Ballesteros D, Kuo M-C. Dry powder compositions having improved dispersivity. WO patent 0132144. 2001.

23. Chew NYK, Chan H-K. Effect of amino acids on the dispersion of spray dried cromoglycate powders as aerosol. In Dalby RN, Byron PR, Peart J, Farr SJ, editors. Respiratory Drug Delivery VIII. Raleigh, NC: Davis Horwood International; 2002. pp. 619–622.

24. Kussendrager KD, Ellison MJH. Carrier material for dry powder inhalation. WO patent 0207705. 2002.

25. Woolfe AJ, Zeng XM, Langford A. Method to produce powders for pulmonary or nasal administration. WO patent 0113885. 2001.

26. Sakagami M, Kinoshita W, Sakon K, Sato J, Makino Y. Mucoadhesive beclomethasone microspheres for powder inhalation: their pharmacokinetics and pharmacodynamics evaluation. Journal of Controlled Release 2002;80:207–218.

27. Dellamary LA, Tarara TE, Smith DJ, Woelk CH, Adractas A, Costello ML, et al. Hollow porous particles in metered dose inhalers. Pharmaceutical Research 2000;17:168–174.

28. Edwards DA, Batycky RP, Johnston L. Highly efficient delivery of a large therapeutic mass aerosol. WO patent 0195874. 2001.

29. Weers J. Dispersible powders for inhalation applications. Innovations in Pharmaceutical Technology 2000;1:111–116.

30. Maa Y-F, Prestrelski SJ. Biopharmaceutical powders: particle formation and formulation considerations. Current Pharmaceutical Biotechnology 2000;1: 283–302.

31. Chickering DA III, Keegan MJ, Randall G, Bernstein H, Straub J. Spray drying apparatus and methods of use. US patent 6223455. 2001.

32. Watanabe W, Ahonen P, Kauppinen E, Jarvinen R, Brown D, Jokiniemi J, et al. Inhalation particles. WO patent 0149263. 2001.

33. Watanabe W, Ahonen P, Kauppinen E, Jarvinen R, Brown D, Jokiniemi J, et al. Novel method for the synthesis of inhalable multicomponent drug powders with controlled morphology and size. In Dalby RN, Byron PR, Peart J, Farr SJ, editors. Respiratory Drug Delivery VIII. Raleigh, NC: Davis Horwood International; 2002. pp. 795–797.

34. Lähde A, Raula J, Kauppinen EI, Watanabe W, Ahonen PP, Brown D. Aerosol synthesis of inhalation particles via a droplet-to-particle method. Particulate Science and Technology 2006;24:71–84.

35. Morton DAV. Method of manufacturing particles. WO patent 0187277. 2001.

36. Mumenthaler M, Leuenberger H. Atmospheric spray-freeze drying: a suitable alternative in freeze-drying technology. International Journal of Pharmaceutics 1991;72:97–110.

37. Maa Y-F, Nguyen P-A, Sweeney T, Shire SJ, Hsu CC. Protein inhalation powders: spray drying vs spray freeze drying. Pharmaceutical Research 1999;16:249–254.

38. Maa Y-F, Nguyen P-A. Method of spray freeze drying proteins for pharmaceutical administration. US patent 6284282. 2001.

39. York P, Hanna M. Particle engineering by supercritical fluid technologies for powder inhalation drug delivery. In Dalby RN, Byron PR, Farr SJ, editors. Respiratory Drug Delivery V. Buffalo Grove, IL: Interpharm Press; 1996. pp. 231–240.

40. Sievers RE, Sellers SP, Clark GS, Villa JA, Mioskowski B, Carpenter J. Supercritical fluid carbon dioxide technologies for fine particle formation for pulmonary drug delivery (Abstract MSDI-173). In 219th ACS National Meeting, San Francisco, CA, March 26–30, 2000. Washington DC: American Chemical Society; 2000.

41. Foster NR, Regtop HL, Dehghani F, Bustami RT, Chan H-K. Synthesis of small particles. WO patent 0245690. 2002.

42. Bustami RT, Chan H-K, Dehghani F, Foster NR. Generation of micro-particles of proteins for aerosol delivery using high pressure modified carbon dioxide. Pharmaceutical Research 2000;17:1360–1366.

43. Bustami RT, Chan H-K, Dehghani F, Foster NR. Recent applications of supercritical fluid technology to pharmaceutical systems. Kona: Powder and Particle 2001;19:57–70.

44. Chan H-K, Kwok PCL. Production methods for nanodrug particles using the bottom-up approach. *Advanced Drug Delivery Reviews* 2011;63:406–416.

5

Methods for understanding, controlling, predicting, and improving drug product performance

David A.V. Morton

Monash Institute of Pharmaceutical Sciences, Monash University, Melbourne, Australia

5.1   Introduction

5.1.1   The complexities and challenges of aerosol performance

Arguably, aerosol drug-delivery systems (inhalers) represent the most complex of medicinal products. Understanding the scientific principles and mechanisms that determine the performance of the aerosol clouds generated is particularly important in their design. It is also critical if intelligent design is to be used in improving their delivery. This chapter introduces some core concepts in aiding this understanding and in providing metrics concerning inhalers' potential performance and their design space during development. The intention here is not however to describe the standard "in vitro" pharmacopeia end-product quality-control impactor and unit-dose tests (see Chapter 6), but rather to outline the constituent parts, and their role in product performance. We start by looking at the context of the performance challenge itself.

The challenge ahead

There is no question that pharmaceutical companies are scientific pioneers; they are developers and operators of leading-edge technologies. Indeed, the 20th century has

Inhalation Drug Delivery: Techniques and Products, First Edition. Paolo Colombo, Daniela Traini, and Francesca Buttini.
© 2013 John Wiley & Sons, Ltd. Published 2013 by John Wiley & Sons, Ltd.

seen some remarkable developments in medicinal chemistry (see [1] for several good examples), and the 21st century is predicted to yield a new era in medicines, resulting from new biopharmaceutical innovations. Many of these products of biotechnology have the potential to be used in inhaled form using dry powder or liquid aerosols (see for example [2,3]).

Our understanding of the basic phenomena underlying aerosolization, powder behavior, and particle–particle interactions is growing, and there are many new and exciting process technologies emerging, offering for example new generations of particle constructs, or new engineered device systems. Yet as particle science and technology have advanced, methods used in bulk formulation (especially in powders) processing and manufacturing methods have in many areas been argued to have lagged behind (for example [4]). There is a famous quotation from September 2003, when the *Wall Street Journal* published a front-page article claiming that pharmaceutical "manufacturing techniques lag far behind those of potato-chip and laundry-soap makers" [5]. Furthermore, in 1996, leading pharma aerosol scientists Richard Dalby and Anthony Hickey stated, "The forces governing dispersion are well documented and consist mainly of electrostatic, van der Waals, and capillary forces. Knowing these forces exist has **not facilitated aerosol generation to any extent**" [6].

Looking specifically at powder technology, this plays a key role in the development and production of most medicines (not just inhalers). It is a cornerstone of many pharmaceutical development projects, from tablets and capsules, to injectables, oral suspensions, and topical products. The approach to powder technology which runs through so many aspects of current pharmaceutical manufacturing processes has in many cases been left virtually unchanged for decades. This can be said of many of the standard primary and secondary processes involved, such as comminution, blending, agglomeration, and coating. It is certainly true of many processes used to manufacture inhaled powders. Despite a brief period when inhaled insulin (Exubera; see for example [2,3,7]) was on the market as a notable exception, being a spray-dried powder, currently there are few significant particulate inhaler products for sale in which the drug is not milled by a micronizer—a traditional process of mechanical destruction, rather than sophisticated particle engineering.

It is highly valid in this chapter, which explores performance and improvements, to ask why this is so, and what are the prospects for improvement.

First, it could be said that current inhaled medicines for local therapies such as asthma and chronic obstructive pulmonary disease (COPD) generally enjoy such wide therapeutic windows and safety profiles that high efficiency and consistency in performance, while very desirable, have not always been demanded.

Second, this chapter argues that consistent with the Hickey/Dalby view from 1996, there is still generally a poor appreciation of the fundamental processes involved in bulk particle behavior and aerosolization processes. Powder/particle technology is not just complicated, it is a highly complex multifactorial phenomenon. It is seen by many as a skilled art, reflecting the often poor understanding of the fundamental science behind it. A better recognition of this, and a resulting approach to understanding, is key to improving product performance levels. Many product and process aspects have been left unchanged due to the attitude, "if it ain't broke, don't fix it."

One factor which has allowed powder technology to slip behind progress made in other scientific disciplines in the pharma field appears to be the traditional/historical structure of academia. Powder/particle technology is a multidisciplinary area, as well as a most complex and challenging applied discipline. Consequently, it has too often fallen between the traditional disciplines taught and driven by academic centers. Use of the available technical knowledge requires a cross-disciplinary approach, and the ability and willingness to cross boundaries.

Where specialist subjects have existed more strongly—and colloid science is perhaps the best example of this in our context—a number of problems exist. Many colloid experts will correctly argue that particle–particle interfaces, interactions, and forces are now well understood and can be readily modeled and explained. However, this understanding has generally been developed using idealized model systems, including near-perfect hard spheres of idealized materials such as silica. Real inhalers contain real drug and excipient particles, which often have an apparently infinite range of surface variations (morphologies, chemical groups, structural order, etc.). Recognizing this is critical in appreciating the challenge ahead for the design and prediction of aerosol inhaler performance.

Third, there is a resistance to change in manufacturing. The regulatory environment has historically led to a largely empirical demonstration of conformity to specification and consistency for a series of production batches. While this has generally been effective, it is not especially technically demanding. One can also argue that in consequence of these effects, the refinement and uptake of new particle or device designs and new processing methods has been slow. Perhaps this will change in the near future with current initiatives such as Process Analytical Technology (PAT) and Quality by Design (QbD) [8].

5.1.2 Understanding powder/particle characteristics: implications for aerosol product performance

So, we know that formulating a drug for practical, efficient, and reproducible pulmonary delivery as particles from a device into a patient is a major challenge. At the core of the problem is the inherent cohesion of drug particles that are physically small enough (aerodynamic diameter of about 1–6 μm or even smaller for systemic delivery) to be efficiently transported as an aerosol into the lung with an inspiratory airflow, especially when they are administered as a powder form from dry powder inhalers (DPIs). Added to this challenge is the "real-world" variation of each and every new material (as noted in Section 5.1.1).

Considering the entire aerosolization process of a powder from an inhaler device during inhalation, it is logical to view it as two critical events. First, the powder should be resuspended and carried from an inhaler device by airflow, termed "entrainment." Second, the powder should be deagglomerated into fine particle form during the aerosolization, often termed "dispersion" or "deagglomeration." The external factors, such as airflow rate and inhaler design (see for example discussions [9–11] and references therein), have been show to be major influences. However, it is important to recognize that not only is our fundamental understanding of aerosolization and real particle interactions limited, but studies on the effect of intrinsic fine-particle cohesion on aerosol performance are not common and are rarely even close to being able to explain observed variation, let alone predict it. Such cohesion confers poor flowability, poor fluidization, and poor dispersion characteristics on powders, which are unable to provide efficient and consistent drug-delivery performance.

Consider the most commonly adopted approach to this issue for DPIs: creation of a homogeneous interactive mixture of fine drug particles and coarse carrier particles for powder inhalers. This interactive-mixture approach has arisen largely through experience, as an empirical solution. The coarse carrier is used to enable the formulation to flow, and powder flowability is achieved in order to meet manufacturing demands, ensuring dose accuracy and dose uniformity, and to enable the formulation to fluidize and be entrained from the inhaler device. However, the drug-delivery efficiency of these mixtures is generally low, since most drug particles are unable to detach from carrier surfaces during the aerosolization phase, due to the strong adhesion between drug and carrier particles [12,13].

Further, a ternary DPI-system approach has been empirically evolved by adding fine additives (in most cases, finely milled lactose powders) into drug–coarse carrier mixtures, leading to an improved drug-delivery performance [14]. Several theories have been proposed which try to explain the mechanism of this ternary DPI system. One is that fine additives occupy the active binding sites on the carrier surfaces and thus reduce the adhesion between drug and carrier particles; another is that the formation of agglomerates of drug and fine additive particles can reduce the effective adhesion between drug and carrier particles, as well as reduce the agglomerate strength. However, the true mechanism is far from clear and is undoubtedly non-trivial (see for example [14,15]).

5.1.3 Liquid systems

One can argue that the basic processes of liquid atomization are much better understood, especially in terms of control, than those of powder dispersion. This understanding of spray processes results from the significant effort invested in developing and predicting liquid-spray behavior in fields varying from fuel injection to instrument design, from printers to spray dryers [16]. The fruit of this knowledge is a growing number of new, standalone, liquid-atomizing inhaler devices.

The aerosol science of droplet atomization and transport still provides highly complex challenges. With different atomization mechanisms, parameters such as surface tension and viscosity vary in influence, but perhaps of greater concern are the differences between the behaviors of solutions and suspensions. As with powders, the interparticle interactions in colloidal suspensions are a prime source of uncertainty in performance. The recent emergence of nanosuspensions during atomization—and their variations—also has important implications for behavior.

Once produced, an aqueous liquid aerosol is an inherently unstable beast, in which phenomena such as evaporation and coalescence feature. Such phenomena should be considered in designing any inhaler.

Despite this better understanding and control of liquid aerosolization, liquid-based inhalers are generally on the decline. The reasons for this are many and complex, but include company marketing strategies, the outcome of Montreal Protocol-driven changes to propellant use, the limit on dose ranges possible in liquids, and the inherently lower chemical stability of many molecules in solution rather than in dry powder forms.

5.1.4 Summary

The purpose of this introductory section has been to identify the many challenges facing the product development of an inhaled form: a challenge often under-estimated by new entrants into the field, usually at their peril. It can be argued from the examples in this section that advances are not obviously led by a sophisticated understanding of fundamentals. However, the limitations of our knowledge should be qualified by the fact that many highly effective, well-produced, and commercially very successful products clearly exist. Consequently, we have found ways in which these issues can be addressed. Many robust techniques and approaches are available, and these are evolving to characterize specific physical properties of the formulation components, the process methods, the devices and their surfaces, the geometry and hydraulic fluid properties, and of course the nature and behavior of the finished product.

In the rest of this chapter, we will briefly introduce a number of the current main approaches used to characterize specific physical and chemical properties of a formulation. With new analytical technologies appearing regularly, this area is continually evolving. Properties of interest include particle and aerosol size distributions, particle forms, amorphous/crystalline content, surface characteristics, and interparticulate cohesion forces. In addition, we shall introduce some material and process strategies for performance improvement.

5.2 Particle sizing

The key parameter for efficient delivery is aerodynamic particle size. The background to particle size distributions, moments, and representations, and their implications, is detailed elsewhere, but the reader should keep at the forefront of their mind the fundamental importance of Stokes' law, and especially the influence of aerosol aerodynamic diameter [17].

No single particle-sizing technique is suitable for all tasks. The particle-sizing approach that is to be used at any stage needs to be selected with care and attention if we are to understand the desired outcome and the nature of the material. The basic techniques will be introduced here in outline only, as a good number of excellent and comprehensive texts exist in this area [18–21].

5.2.1 Sieve analysis

For larger particles, notably carrier particles used in DPI formulations, sieve analysis is often used. The powder is introduced on to a stack of analytical sieves, which are arranged vertically in order, with the largest aperture sieve at the top, then the next largest, and so on. The stack is then mechanically vibrated, in order to encourage the powders to be exposed to each aperture: particles that can geometrically fall through the mesh will continue to descend until they are trapped on the stage below the smallest aperture they can pass through. The powders are vibrated for a validated time, beyond which further changes are minimal. The powder masses retained on each stage are then determined.

In practice, particle sizing with a sieve is limited by the ability to deagglomerate particles, so sieving below approximately 50 μm is often impractical because the cohesive forces are greater than the gravimetric detachment forces. Wet-sieving or air-jet-sieving can in some cases allow smaller particles to be effectively sieved.

5.2.2 Image analysis

Particle size (and form) may be assessed visually under a microscope. The microscope method involves preparing suitable particle dispersion for inspection, where preferably all particles are separated and can be independently distinguished, and then comparing the projected 2D areas of a large number of irregularly shaped particles to a standard graticule or series of circles of known diameter.

Determination requires considerable operator skill and is tedious. However, it is important to note that this is the best approach discussed here for attaining information about particle shape. While only two dimensions of the particle can be seen at any one time, the shape information allows detection of particle habit and general morphological issues, and of aggregation.

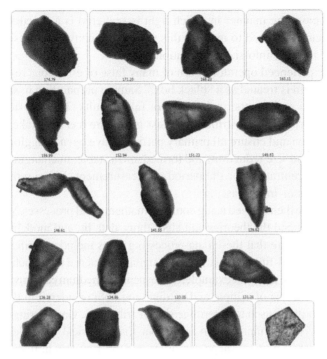

Figure 5.1 Lactose samples imaged using the Malvern Instruments Morphology G3 system. Image courtesy of Monash University-using Morphology G3 instrument

Optical-microscope methods are limited by the microscope's resolving power, and hence are practical only for particles in the order of tens of microns. For micron- to submicron-sized particles, an electron microscope can be used.

Computerized image analysis can be used to automate and overcome some of the operator issues. However, this approach requires software that is intelligently defined and can decide, for example, where the edges of each particle are and make assumptions about the particle thickness.

Computerized microscope optical image analysis can provide a host of shape-related data, as indicated here for lactose samples, where for example circularity, elongation, convexity, and so on can be measured (Figure 5.1).

5.2.3 Light scatter

Light-scattering techniques are currently the most popular systems in use for particle sizing of micronized drug particles, fine or larger excipients, and composite formulated systems. These systems function by measuring the quantity of light scattered by individual particles as they pass through a beam of intense light (often a laser). Instruments may use principles of Fraunhofer diffraction and Mie theory (or Doppler velocimetry) in order to derive particle size information from the scattered light. In each case, the manner in which light is scattered is a function of particle size, and detectors are used to monitor this scatter and convert these signals, using mathematical models, into size distributions.

However, experienced operators warn of many false or erroneous results that can occur if the system is treated as a "black box." Sample preparation in such particle-sizing methods is perhaps the main cause of false results. These systems will not provide a true particle size distribution unless great care has been taken to validate sample preparation and ensure all primary particles have been deagglomerated (but without milling and size reduction) and formed into a stable, unchanging suspension of a suitable concentration for the period of measurement. Measurement samples may be in liquids or in air/gas.

Particle sizes are then derived using complex mathematical processes. While it is not essential to understand the theories and algorithms, data fit, and modeling used, it is important to appreciate that these data-processing steps include significant levels of assumption and model-fitting activities. Poor signal or material data, such as refractive indices, can yield false data. For example, the suspension medium can provide problems if the refractive indices of sample and medium are not considered in the context of the specific algorithms used to determine size distributions. Considering particle size distributions, it is also important to remember that spherical equivalent diameters are normally quoted; particle shape can differ widely from this in reality. Similarly, these techniques often provide volume equivalent diameter rather than aerodynamic diameter—all of which must be appreciated by the user in data interpretation.

5.2.4 Time-of-flight

Aerodynamic size can be measured in real time by a time-of-flight aerosol sizer, where particles are separated and accelerated through a well-defined nozzle and their flight between two portions of a split laser beam is timed. The flight time can be directly used to calculate an aerodynamic equivalent diameter. Again, key limitations of this light scatter-based real-time instrument include the sample preparation, interpretation/algorithm issues, and presentation/representation of the data.

5.2.5 Other methods

Many other methods exist in the arena of particle sizing. Sedimentation has arguably become less popular in recent years, probably due to the advent of modern computer-controlled light-scatter instruments. Similarly, the Coulter-counter method and related electrical sensing methods have become less popular. Despite this apparent fall in popularity, all these methods are useful and may find niche value in specific requirements.

Table 5.1 Preferred methods of particle sizing

Size range	Direct measurement	Indirect measurement
Carrier particles above ~50 μm	Optical microscopy	Laser diffraction Time of flight Sieve analysis
Carrier particle fines below ~50 μm	Electron or optical microscopy	Laser diffraction Time of flight
Drug agglomerates/pellets above ~100 μm	Optical microscopy	Laser diffraction
Micronized drug ~0.1–10 μm	Electron microscopy	Laser diffraction Time of flight
Nanomilled drug ~0.01–1 μm	Electron microscopy	Photocorrelation spectroscopy

Apart from traditional impactors and impingers familiar to pharmaceutical aerosol science, various other inertial separation methods are available, such as the electrical low-pressure impactor (ELPI), the Stöber centrifuge, and cyclone trains. Again, while less common, these methods can certainly offer some unique benefits.

For submicron particles—a size range not often of interest to inhaler systems—a number of specialist methods are available, from condensation nuclei counters to photocorrelation spectroscopy (e.g. the Zetasizer from Malvern Instruments, Malvern, UK), and including recent alternatives such as NanoSight technology (NanoSight Ltd, Wiltshire, UK).

Shape measurement has been an area of continuing interest, and instruments have used both image analysis (such as the Morphologi G3; Malvern Instruments, Malvern, UK) and changes in light scatter pattern (e.g. ASPECT from Biral Ltd, Bristol, UK).

In practice, particle sizing conducted by pharmaceutical scientists developing inhaled products tends to be limited to a small number of the most common techniques, using direct visual measurement or indirect derivation measurement. Preferred methods are listed in Table 5.1.

5.3 Powder and particulate characterization systems

5.3.1 Introduction

The physicochemical issues influencing powder product performance have been outlined in other chapters and illustrated in the introductory section of this chapter, including (but not limited to):

- How bulk powder and individual particles behave during primary and secondary processes in manufacture.

- How practical it is to meter the device: including flow and segregation.

- Behavior in storage and transport, including chemical and physical stability.

- Performance in the device: including flow and deagglomeration.

Other physicochemical factors also have significance for inhalation, such as the taste of a deposited powder and other sensations stimulated, such as cough. Related to this, the dissolution rate is also an issue, at specific points in the respiratory tract for example, and despite arguably receiving little attention, has implications for drug bioavailability.

Consequently, intrinsic powder physicochemical properties provide the basis for product performance. These properties include: size distribution, surface area, crystallinity, purity, moisture, electrostatic charge, shape factors, surface morphology and chemistry, and density. Size distribution has been treated in its own right in Section 5.2. This section looks primarily at interfacial issues.

5.3.2   Powder cohesion and adhesion

The primary issue under consideration here is powder cohesion. Most inhaler powder formulations contain both drug and excipient particles, and hence we are concerned with both "cohesive" (drug–drug) interactions and "adhesive" (drug–excipient) interactions. While excipient–excipient "cohesive" interactions may be important too, focus tends to be on interactions directly concerning drug. The bulk behavior of powder mixtures is a complex function of the adhesive and cohesive forces present.

It is important to consider the nature of such forces. These include the ubiquitous van der Waals forces, as well as possible influences from electrostatic and capillary forces. Several textbooks provide a good fundamental background to this area [12,13].

The forces of adhesion and cohesion experienced by the drug particles should not be too strong, as the powder may flow too poorly to be mixed, in order to be transported and filled during manufacture, and may not be released from the device, or else drug particles may not detach during inhalation. If such forces are too weak, powders may segregate at critical stages, resulting in unacceptable variation in dosing. In addition, it is possible that too-weak forces may also be detrimental to aerosol quality in some poorly matched systems [22].

So, we need to be aware of how the powder physicochemical properties may influence powder cohesion/adhesion, and how we can assess these properties.

Particle size is very important: as particle size decreases, so surface area and surface energy increase. In simple terms, the greater the area of contact between two surfaces, the stronger the interactions. Particle size analysis was discussed in

Figure 5.2 Tabular and needle particles of milled salbutamol sulfate (left) pack together differently to spray-dried particles containing salbutamol sulfate of a similar size (right). Image courtesy of Monash University-using Morphology G3 instrument

Section 5.2. However, it is important to note that a mean particle size value alone may not tell the complete story, as particle size distribution and shape factors can also play a major part in powder cohesion/adhesion. For example, two powders with similar measured mean sizes may change in cohesion with spread of the size distribution. Particle shape is often ignored when size distributions are measured and, for example, tabular or plate-shaped particles or needles may flow less well than more anisotropic particles, due to differences in the alignment of adjacent particles, packing efficiency, and the area of contact points (Figure 5.2).

5.3.3 Microscopic material characterization

In this section, a range of techniques will be introduced that allow characterization of material properties, and especially surface properties, as these are the key interfaces dictating performance [12,13].

Surface-area measurement plays an important role in this context, and can help in quality-control procedures. Specific surface area is measured as the surface area per unit volume or surface area per unit weight. Cohesion/adhesion may be dependent on the surface area present, but are also strongly dependent on the physical and chemical nature of that surface. Several techniques are available to provide information in this area.

The amorphous/crystalline content of each particle surface can be critical. In general, an amorphous surface is less stable and higher in energy than a crystalline surface. With the impacts experienced during milling or mixing, many otherwise crystalline particle surfaces can have amorphous regions induced upon them.

The hardness of a surface can be important, as the greater deformation of a softer material on contact with another surface can lead to increased contact area and stronger interactions.

The chemical groups exposed at particle surfaces can have strong influences on interactions. The range of van der Waals forces possible can change substantially depending on the polarity and polarizability of such groups exposed. Similarly, the tendency to generate charges due to tribological interactions can therefore influence cohesion/adhesion.

The presence of adsorbed moisture on surfaces has the potential to enhance interactions, via hydrogen bonding. If sufficient water is present to form capillary bridges, the cohesion/adhesion of a powder can be very much increased, and for particles in close contact, such interactions may be very much greater than van der Waals or electrostatic forces. Consequently, the presence of moisture in a powder can be catastrophic for powder inhaler formulations.

Surface roughness is capable of influencing cohesion/adhesion. No simple rule exists to predict the effect of roughness, and it is a very difficult property to quantify usefully. Surface roughness at a very small (nano) scale may reduce cohesion/ adhesion if mechanical protrusions limit the area of contact between two approaching surfaces. Alternatively, surface roughness at a micro to macro scale may in some cases lead to increased surface contact area via mechanical interlocking or mechanical trapping of small particles in larger fissures (Figure 5.3).

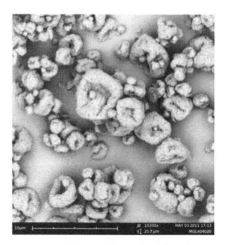

Figure 5.3 Spray-dried area showing particle trapping within surface wrinkles. Reproduced with permission from Tomás Sou, Lisa M. Kaminskas, Tri-Hung Nguyen, Renée Carlberg, Michelle P. McIntosh and David A. V. Morton; The effect of amino acid excipients on morphology and solid-state properties of multi-component spray-dried formulations for pulmonary delivery of biomacromolecules; European Journal of Pharmaceutics and Biopharmaceutics-in preparation

In addition to these aspects, particle density has recently become an area of substantial interest in pulmonary delivery formulation. It is known by environmental and occupational health aerosol scientists that the respiratory problems associated with soot/combustion particles can easily be underestimated: due to the fact that they often exist as complex chain or porous agglomerates, these particles appear to have relatively large physical diameters when measured by light-scattering techniques. However, this measurement method gives a spherical equivalent diameter of the envelope around such particles, but they are in fact highly porous; if their aerodynamic diameter is measured, they appear much smaller, as they are in effect very low-density, and hence have a much greater respirable nature. This concept has been adopted by inhaler formulators, by deliberate engineering of low-density particles [23]. The advantages are simple to understand: as for highly porous spheres with a large physical diameter, the contacts between particles per unit volume are much reduced. Consequently, they may flow and be deagglomerated in a manner analogous to relatively large-sized particles. However, they can behave aerodynamically once suspended as very small particles, and hence fly deep into the respiratory tract.

Managing the interfacial particulate forces, for example via particle-engineering routes, is seen as central to successful DPI formulation. Consequently, much attention has been focused on producing "smart" formulations, such as a low-density option. It is also interesting to note that many of the low-density particles so far engineered contain some component of lipids in their formulation—and it is possible that such components could well chemically reduce surface contact cohesion in addition to the physical surface contact shape and density factor.

So, controlling the particle surfaces themselves has become a major focus (both in powders and in liquid suspensions). A popular option is to produce crystalline surfaces with a minimum of amorphous surface content. In the simplest form, many formulators take efforts to minimize amorphous content produced by primary (milling) and secondary (mixing) processes by conditioning the powders, via storage at controlled humidity [24]. This is intended to encourage any amorphous material to recrystallize in a controlled manner, rather than create solid bridges [25].

A second option is to focus on generating highly crystalline particles. Such has been the focus of advanced precipitation techniques such as those which use supercritical fluids (SCFs) [26]. One observation from this approach is that while greater stability and consistency can be achieved, the perfect and flat crystal faces generated may ironically not be ideal, as they may allow areas of contact between particles to be significantly enhanced, making powders more difficult to aerosolize.

A further approach has been to produce highly smoothed carrier surfaces, for example those of lactose carriers, via controlled or partial recrystallization [27]. Such surfaces show improved performance where adhesion is reduced, but if there is too little adhesion, segregation may become a problem.

Arguably the most popular approach to the interface problem has been to produce very thin coatings on particles. The ultimate idea here is to make all particles look "the same" from the exterior, with a controlled low surface energy. The main initial work in this area was conducted on larger carrier particles, as these are less of a challenge to coat. Greater effectiveness however has been seen when coating micron-sized drug particles, and very significant performance improvements have been demonstrated. Magnesium stearate has been the most popular coating material, but other materials with lubricant or surfactant properties have included leucine and phospholipids [27].

Two coating methods have been used: co-milling and co-spray-drying. In the former, techniques including jet-milling, ball-milling, and mechanofusion have been used [27]. For the latter, co-spray-drying with the amino acid leucine has been explored to good effect [28]. Similarly, various proteins, such as albumen, have shown benefit as added excipients. Phospholipids such as DPPC have also been used in this context. In each case, a key factor seems to be surfactant self-assembly occurring preferentially at the surface during drying, which provides a uniform low-adhesion coating.

Surface measurement techniques

Surface-area measurement techniques are well reviewed in texts such as Allen (1997) [29] and Webb and Orr (1997) [30]. The most common method for measurement of specific surface area examines the number of nitrogen gas molecules adsorbed as a single layer on the surface of a known weight of powder (BET (Brunauer, Emmett, and Teller) method). The powder is first weighed in a vacuum, then reweighed several times under increasing nitrogen pressure. Nitrogen is adsorbed on the surface, thus increasing the mass of the powder.

A plot (or isotherm) is constructed from the weight of adsorbed N_2 per gram of powder against pressure. Mathematical methods are then used to determine the number of N_2 molecules which form a monolayer per unit. As a nitrogen molecule has known cross-sectional area, the surface area can be calculated. This method requires correcting equations (BET isotherm equation) for the fact that the gas forms more than one layer on the particle surface. As well as surface area, the principles of this technique can be extended to look at pores in surfaces and microporosity. As an alternative, mercury intrusion can be used to study surface pores [29].

Inverse gas chromatography (IGC) is a technique used to probe the surface energy of a powder [31]. IGC can be regarded as the reverse of traditional analytical gas chromatography (GC), as the roles of the stationary (particulate solid) and mobile (gas or vapor) phases are inverted. Whereas for GC, a standard column is used to selectively separate the analytical sample gases of interest, for IGC a standard probe molecule is injected into a column packed with the powdered solid sample under analysis. The retention time of each probe molecule is quantified and used to characterize specific aspects of the surface under investigation. In principle, IGC can yield information on the surface-energy properties of solids, such as the polar and dispersive surface energies, including information on the crystal planes or amorphous content exposed. In practice, IGC measurement requires great care, and interpretation can be problematic. One limitation is the current prevalence of measurements at "infinite dilution," where only small amounts of probe molecule are injected (to allow negation of any conflicting probe–probe interactions), but the technique in this case tends to yield information on only the highest-energy sites, which may not reflect the bulk of the material surfaces, and attempts have been made to extend to a "finite dilution" [32].

Alternative methods for studying surface energies include the measurement of contact angles for specific liquids and the measurement of capillary intrusion into a powder column [33]. In each case, a range of liquids must be selected, with different polar/nonpolar characteristics, but these liquids should be nonsolvent for the material under examination.

Dynamic vapor sorption (DVS) is a technique used to probe surface physical state: it looks at the uptake rate of a solvent on to a sample surface. Vapor in varying concentrations is passed over the analyzed surface and the resulting change in mass is measured [34,35]. For DPI powders, this technique is often used to probe water-vapor sorption in order to investigate surface amorphous contents (other solvents may also be used to probe additional properties), as well as to probe solvate formation.

An alternative method for studying amorphous content is the use of calorimetry, notably microcalorimetry [33]. However, this technique probes the bulk of each particle, rather than focusing on the surface state.

The atomic force microscope (AFM) is a technique now used widely for a number of characterization roles in DPI materials. The colloid-probe technique allows the quantification of the total interaction force between an individual drug particle and a substrate surface. In this process, a drug particle may be attached to the probe tip, and the force of adhesion on to a probe surface analyzed directly. This technique is limited where the area of contact is not known, but the cohesive–adhesive balance (CAB) approach was developed to overcome this issue [36]. The AFM can also be used to examine surface roughness at a resolution that approaches atomic scale. It may further be used to assess material hardness, and may be adapted to reveal some calorimetric properties or derivatized to probe chemical interactions.

Several methods are available for probing chemical species at the surfaces of particles. Many electron microscopes have elemental detections, such as the EDAX (energy-dispersive X-ray spectroscopy), a technique employing X-ray radiation for compositional analysis, which can be used to indicate chemical (if not molecular) composition to supplement images. The techniques of XPS (X-ray photoelectron spectroscopy; also known as ESCA) and TOFSIMS (time of flight secondary ion mass spectroscopy) may also be used to probe surfaces: the former gives elemental information but some structural details from the electronic structure of each element (for example, the electronic bonding identity of a carbon atom); the latter gives more specific structural details in the form of mass spectra (e.g. [37]). Another alternative is the use of Raman spectroscopy. A Raman probe or other analytical method can be added to a microscope to reveal information about surface or composition species [38].

The standard technique for measuring material water content is the Karl Fischer technique. Alternatively, a loss-on-drying approach can be used. In each case, the technique cannot easily distinguish surface water from bound water—and this is clearly critical in cohesion issues.

Summary

These specialist analytical tools generally require substantial skill, both to obtain valid data and to allow for effective interpretation.

The data derived must also be regarded with care, as no one technique provides all the answers: for example, rugosity, surface energy, and moisture content are all valuable parameters, but there is very little evidence that these data in isolation provide any improved ability to predict how powders will perform, though they may help indicate where problems have occurred. So the outcomes of any powder characterization and testing in isolation need to be treated realistically: it is very important to review and understand the importance and role of each of these measurements, in order to ensure a high degree of skill in their selection, measurement, and interpretation.

These measures may best be regarded as providing core value as part of a practical powder-quality management system, for example in process control (generally PAT or QbD [8]). A critical point is surely that control of the input particulate materials from suppliers is vital, and lies at the core of achieving consistently successful primary powder materials.

This quality-management system should then extend to control of the primary and secondary process methods used to create and then assemble a formulation for use in the medicinal DPI product.

5.3.4 Methods for studying bulk powders

A variety of methods for the assessment of powder flow have been developed. Of these, powder shear testing is arguably the most widely used [39]. The force required to shear a powder under well-defined conditions is measured. This area has a substantial background, and was pioneered by Jenike, who also developed the theoretical framework. Shear testers range in design from ring shear testers to uniaxial shear testers.

Flow characteristics are also tested by assessing powder density and compressibility. Poured and tapped bulk densities are included as pharmacopeial tests, and these values can then be expressed using Carr's index or the Hausner ratio [40]: the ratios between these two bulk densities. Other tests [41] included in the evaluation of flow are the angle of repose, the angle of spatula, and the ability or time taken to flow out of an orifice of specific diameter. The repeat time to avalanche of a powder rotated in a drum can also be used to characterize powder flow [42].

A recent approach, which shows perhaps the greatest promise, is the development of powder rheometry [42,43]. One of the main limitations of the tests listed in this section is the notorious variability between them, despite attempts at standardization. It is easy to appreciate why this is when one considers the view of Geldart: that powders uniquely exhibit mixtures of features otherwise associated with either solids, gases, or liquids, in that, unlike simple states of matter, powders can be made to deform, can be compressed, and can be made to flow: a powder behavior is highly dependent upon and variable with its history [4]. The shear testing approach makes attempts to overcome this conditioning problem by consolidating the powder under a significant load. A criticism of this in the context of inhalation is that this history is unrealistic, as few powders in an inhaler will be compressed like this. The recent Freeman powder rheometer [43,44] allows a series of tests of powder resistance to a helical blade under a range of preconditions, both consolidated and with a range of aeration states.

Alternative methods for assessing deagglomeration

The potential benefit of developing a dispersibility screening tool, in the form of flow titration, is that it overcomes the many disadvantages of conventional impactor testing for very-early-stage assessment of the aerosolizability of engineered drug particles [45]. Cascade impactor tests are notoriously time-consuming (both in operation and in subsequent chemical analysis), are flow rate-limited, are of low resolution, and are generally unable to provide size detail above 10 microns. They can also be unreliable and more operator-dependent than other forms of particle-sizing.

Many different measures of powder dispersibility (as alternatives to impactors) can be found in the literature. Such tests have been considered inappropriate for testing many DPI dispersions, where there has been a need to distinguish chemically different dispersed particles, for example lactose particles from drug particles. Real-time light-diffraction methods appear to be the most appropriate option. Such methods can be made both rapid and reliable, providing a high-degree of dispersion quality detail. For example, systems are available which use the light-scattering approach noted in Section 5.3.3, but which are designed to monitor aerosol size from the inhaler plume as a function of time. Alternatively, photodetectors can be employed to follow the dynamic behavior of particles, as well as their size distributions, using particle image velocimetry and Doppler effects.

High-speed video image capture can be used in this context, and can be very valuable in gleaning information on the behavior of particles emitted from the device. In addition, it has been possible in some cases to make transparent versions of devices in order to follow particle paths within devices, and to learn much about deagglomeration processes [46].

Other methods used to assess bulk powder deagglomeration are similar to those used to study the physical stability and tendency to segregation of ordered mixtures. These include vibrational tests on sieves or in powder stacks, and centrifugal acceleration tests [12,41].

Finally, this section would not be complete without recognizing the benefits of mathematical modeling of such aspects as flows, impacts, and fragmentation of agglomerates. Substantial improvements in mathematical techniques and engineering models now allow us to simulate the behavior of many of the processes going on in our complex systems [47].

5.4 Practical issues in process control

Most formulations consist of mixtures. Hence it is critical to control excipients as well as drugs, and both their form and the state of mixing are important. We have already discussed in this chapter aspects of the material properties that need to be controlled. Having a strong control of the ingredients in a formulation technology is only a step on the road to a successful product. Control of processing is vital, and significant problems can still be encountered when designing systems to match the selected inhaler device.

5.4.1 Common primary and secondary processing methods and issues arising for control

Milling

Mechanical processing (blend formation or particle size reduction) is usually intended to achieve a particular material structure, and relies on imparting energy and particulate motion, including impacts, rubbing, and the making and breaking of material contact points. This movement is associated with various tribological effects, and appears to induce short- to medium-term physical changes in many pharmaceutical materials (drugs or excipients). These in practice are inherently chaotic and not well characterized or defined. Mechanical processing can take the form of a blending or a milling step, and the main difference between these is the level of energy and the manner/rate in which the energy is applied.

Physical changes, such as amorphous disorder in the surface structure of particles, electrostatic charge build-up/change in bipolar charging, or agglomerate formation/rearrangement, have been observed, especially in relation to the higher energy levels of milling. These changes can be of significant interest and importance as they may provide an unacceptable degree of uncertainty, change, and inconsistency in product behavior, potentially leading to catastrophic product failure due, for example, to the formation of solid bridges where recrystallization has occurred at contact points.

Mixing

Blending of the components of a powder inhaler formulation provides a critical secondary step in the creation of a stable, consistent, and effective product. Despite its importance, there is very little published work examining the effect of blending on powder stability and performance. For example, where fine lactose ternary component "fines" are added to improve aerosolization of an ordered mix, two fine components (drug and excipient) must be incorporated. This is not a trivial process to control, given the high cohesion present. A very complex variety of structures can form in many sizes and ratios—which will be dependent on the extent, mode of application, and energy of blending. This will substantially alter the location of the drug particles within the blend, affecting performance—for example, the drug might attach to small particles in complex "multiplets" or to highly active sites [14].

Anecdotal evidence from industrial sources indicates that "over-blending" in ternary DPI formulations often occurs, where the fine particle fraction (FPF) decreases with increasing time/energy, which may be attributed to drug particles migrating and adhering strongly to the carrier surface, where they are then more difficult to remove.

Consequently, control of both primary and secondary processing areas is highly significant, and remains a largely empirical domain. The improving lessons from PAT are likely to yield substantial dividends in this area, especially where it is possible to include on-line control techniques [8]. The benefits will be to reduce risk, by identifying end points and hence providing critical values for process design space.

5.5 Biopharmaceutical powder stability

A brief mention will be made here of the complex world of formulating biological molecules. These categories of agents, from short peptides to large proteins, from nucleic acid fragments to attenuated vaccines, provide a distinct challenge in the area of formulation and production. Biological molecules tend to be substantially more prone to inactivation via both chemical and physical degradation pathways. Consequently, a large body of expertise has developed in the area of producing stabilizing glasses to keep such materials from damage during storage (see for example [48,49]). The primary concern is to produce a stable glass that minimizes exposure of the biological molecule to chemical or physical stress. Much of the work in this area has focused on the development of spray-drying [50]. In analytical terms, techniques are used to study the glass structure (such as X-ray diffraction) and to determine the glass transition temperature (Tg), which indicates the temperature below which the glass remains free of excessive solid-state movement (hence providing greater stability of the biological molecule). Tg is normally obtainable from differential scanning calorimetry (DSC), preferably using modulated DSC to achieve the best resolution and sensitivity.

5.6   Liquids: solutions and suspensions

Liquid-based aerosol devices can be divided into solution-based and suspension-based systems. For solutions, the issues influencing aerosolization are largely dictated by the device mechanism of spray formation. In most cases, this can be influenced by liquid physical properties such as viscosity and surface tension.

In the case of pressurized metered-dose inhalers (pMDI)s, the additional consideration is the vapor pressure of the propellant, especially if blends of propellants are used. The effective energy stored within the liquid as it then boils on exposure to atmospheric pressure and temperature will have an effect on droplet size. The physical processes of droplet formation from propellant-based inhalers are nonetheless relatively well characterized and can be predicted with reasonable confidence [51]. The role of the actuator is also relatively well known, and with current experience, droplet sizes can be predicted to a good approximation from the shape and dimensions of the orifice [52].

As for nebulizers, there are now a wide variety of mechanisms employed to break liquids into droplets at the core of each type of system. Each mechanism applies energy in a specific manner to create a large increase in surface area as the liquid is torn into discrete droplets. Each mechanism used for droplet generation provides its own advantages and limitations, including droplet size, droplet size distribution, rate of droplet generation, velocity of generation, potential damage to delicate molecules during the generation, and so on. The main types are listed in Table 5.2 and a number of comprehensive reviews are available [54–56].

Droplet systems have one major inherent physical disadvantage compared to powder systems: the drug in a liquid system is inherently present in a low concentration. To take advantage of the controlled liquid atomization process in order to get very fine droplets, it is rarely possible to have more than a small percentage of drug either in solution or in suspension. Consequently, dose delivery rate is restricted. Further, when considering the performance of droplet generators, whether they be from propellant-based systems or mechanically generated via a

Table 5.2 Main categories of liquid-atomizing device and formulations. Adapted with permission from Morton DAV, Jefferys D, Ziegler LR, Zanen P. Workshop on devices: regional issues surrounding regulatory requirements for nebulizers. Proceedings of Respiratory Drug Delivery Europe Conference. Davis Healthcare International; 2009. pp. 129–148

Atomization mechanism	Brief description and exemplification
Two fluid	Liquid drawn up and sheared into drops by compressed air; e.g. Pari LC
Liquid pressure	Liquid forced under pressure through a microorifice mesh, forcing droplet ejection, or through microchannels, impinging and fracturing; e.g. Boehringer Ingelheim Respimat
Ultrasonic fountain	Piezoelectric element induces high-frequency capillary waves at liquid surface, which break into drops; e.g. Multisonic
Ultrasonic mesh	Piezoelectric element induces high-frequency vibration, coupled with a microorifice mesh, either directly or adjacent to it, forcing droplet ejection from the mesh; e.g. Omron MicroAIR, Pari eFlow, Respironics I-Neb
Electrospray	High voltage imposed at a liquid filament causes electro-hydrodynamic liquid fracture into drops; e.g. Ventaira Mystic

Formulation principle	Brief description and exemplification
Solutions	Dugs dissolved in a liquid, generally aqueous, system, also optionally containing excipients; e.g. Ventolin Nebules GSK
Microsuspensions	Micron-sized drug solid particles suspended in a liquid, generally aqueous, system, also optionally containing excipients; e.g. Flixotide Nebules GSK
Nanosuspensions	Submicron-sized drug solid particles suspended in a liquid, generally aqueous, system, and generally requiring surfactant

nebulizer, the evolution of each droplet with travel and time is one of the key uncertainties. Droplets, by their nature, are unstable, and in almost all cases will be reducing in size constantly due to evaporation. These changes will be driven by environmental factors outside the control of the patient (temperature, humidity). Consequently, there is an uncertainty over the appropriate aerodynamic size distribution for respiratory system transport and delivery. This means that any droplet size-measurement techniques (see Section 5.2) need to be considered carefully.

5.6.1 Liquid formulation stability

Solution-based systems are also less popular than particulate-based systems due to their inherently reduced chemical stability: the enhanced freedom of movement and exposure to chemical attack for a molecule in solution makes it a greater challenge to maintain chemical stability.

Suspensions may have greater chemical stability but pose challenges to the physical stability of suspension. Suspension-based systems will be prone to factors including sedimentation, flocculation, and caking. Consequently, formulation approaches are required to ensure that any suspension can be resuspended into its desired state just by a simple physical agitation. This often requires traditional suspension formulation approaches such as the use of surfactants or polymers and the adjustment of liquid polarity or density. The science of such colloids is generally less well known for propellants than for aqueous systems. The surface of the container can be important too.

Methods are therefore required to characterize and optimize these properties. Suspension stability can be measured by particle-sizing techniques as well as by turbidity measurements and measurement of caking density/height. Particulate interactions can be assessed, for example using liquid-based AFM [57]. Adsorption studies are also useful in understanding surfactant and polymer interactions, for example using a quartz-crystal microbalance.

Suspensions are prone to partial solubility factors, and repeated solution and precipitation or Ostwald ripening can be a problem.

Nanosuspensions and liposomal formulations are emerging as significant future formulation approaches: given their smaller size, they can behave part way between traditional suspensions and solutions. Methods of characterization will also vary, and particle-sizing techniques in particular must be suitable for such size ranges (see Section 5.2).

5.7 Conclusion

The intention of this chapter has been to provide a brief and general introduction to factors that affect the performance of inhalers, and our scope for monitoring, understanding, and improving these devices. This chapter is certainly not comprehensive. It is worth recognizing that technology is constantly shifting, with new trends in formulation, devices, and instrumentation. For example, an emergent area is the new generation of condensation devices (such as the Staccato device from Alexza Pharmaceuticals, CA, USA [58]) which produce very fine "smokes" and present strong advantages in their own right. Good product characterization is key to their future.

This chapter has contrasted our ability to predict and control, from a mechanistic point of view, using liquid-based versus dry powder systems. The reader may ask why liquid-based systems are not more prevalent. Some of the reasons are provided here, but much is due to environmental, political, and commercial influences. Current patterns indicate that powder systems will continue to grow in dominance.

It is remarkable that much about interparticle and interfacial interactions and how they control bulk behavior remains poorly understood. For example, the basic action of lactose fines remains not just a mystery but often a contradiction. Fine lactose particles are more cohesive than most drug particles for a comparable size distribution. So why is it that adding fine lactose appears to improve aerosolization of the drug? "Active site" theory alone is not enough to explain this. Also, why is it that more cohesive drugs can give higher FPFs? This is highly counterintuitive unless the relationship between a device and the agglomerate structure is considered. Formulations cannot truly be developed in isolation, without considering deagglomeration mechanisms.

Evolution of a blend during mixing also appears to be a highly complex phenomenon. In combination formulations, different drugs change performance at different (even contradictory) stages of mixing. What is going on?

Where adhesive–cohesive bonds form, there is plenty of evidence that relaxation of the powder over hours to weeks leads to changed FPFs (usually decreased). However, it is not clear what is going on. How far can the surface properties of such systems be studied by improved IGC, surface area, AFM, calorimetry, or other methodologies?

So, a basic understanding of the materials, mixtures, and processes is important, and the outcomes and limitations of material characterization and testing need to be recognized in this context. Our current products are primarily based on good and practical quality-management systems and control of the input materials by suppliers and of the primary and secondary process methods used to create and then assemble the formulations, as these factors lie at the core of consistently successful formulations. This quality-management approach should then be extended to the medicinal device and assembled product.

Ultimately, inhaler formulations need to be matched to the device of choice, and the end-product formulation performance in this environment is the key issue. After this control is achieved, a comprehensive and robust pharmacopoeial in vitro test program (including impactor and dose-uniformity tests), including both physical- and chemical-stability tests, remains the key indicator of inhaler product performance.

However, end-product testing is not the key to the future. In quality assurance language, "you cannot test quality into a product." Characterization and understanding are central to technology advancement in the inhalation product arena. What is certain is that the future offers many exciting opportunities for achieving better understanding of these important, complex, and fascinating systems, and the benefits of such improved understanding will yield step changes in our capabilities and new product opportunities.

References

1. Le Fanu JD. The Rise and Fall of Modern Medicine. New York, NY: Carroll & Graf; 2000.

2. Newman S. Respiratory Drug Delivery: Essential Theory and Practice. Boca Raton, FL: Davis Healthcare International; 2009.

3. Smyth HDC, Hickey AJ. Controlled Pulmonary Delivery, Advances in Delivery Science and Technology Series. Berlin: Springer; 2011.

4. Geldart D. Principles of Powder Technology. Chichester: John Wiley & Sons, Ltd; 1990.

5. Abboud L, Hensley S. New prescription for drug makers: update the plants. Wall Street Journal. September 3, 2003.

6. Hickey AJ. Inhalation Aerosols—Physical and Biological Basis for Therapy. New York, NY: Marcel Dekker; 1996.

7. Sadrzadeh N, Miller DP, Lechuga-Ballesteros D, Harper NJ, Stevenson DL, Bennett DB. Solid-state stability of spray-dried insulin powder for inhalation: chemical kinetics and structural relaxation modeling of Exubera above and below the glass transition temperature. Journal of Pharmaceutical Sciences 2010;99(9): 3698–3710.

8. Ermer J, Miller JHMcB. Method Validation in Pharmaceutical Analysis. Veinheim, Germany: Wiley-VCH; 2005.

9. Xu Z, Mansour H, Hickey AJ. Particle interactions in dry powder inhaler unit processes: a review. Journal of Adhesion Science and Technology 2011;25(4–5):451–482.

10. Islam N, Clearey MJ. Developing an efficient and reliable dry powder inhaler for pulmonary drug delivery—a review for multidisciplinary researchers. Medical Engineering & Physics. In press.

11. Behara SRB, Kippax P, Larson I, Stewart PJ, Morton DAV. Insight into pressure drop dependent efficiencies of dry powder inhalers. European Journal of Pharmaceutical Sciences. In press.

12. Podczeck F. Particle–particle Adhesion in Pharmaceutical Powder Handling. London: Imperial College Press; 1998.

13. Zeng XM, Martin GP, Marriott C. Particulate Interactions in Dry Powder Formulations for Inhalation. London: Taylor & Francis; 2001.

14. Jones MD, Price R. The influence of fine excipient particles on the performance of carrier-based dry powder inhalation formulations. Pharmaceutical Research 2006;23(8):1665–1674.

15. Shur J, Harris H, Jones MD, Kaerger S, Price R. The role of fines in the modification of the fluidization and dispersion mechanism within dry powder inhaler formulations. Pharmaceutical Research 2008;25(7):1631–1640.

16. Lefebvre AJ. Atomization and Sprays. Washington, DC: Hemisphere; 1989.

17. Hinds WC. Aerosol Technology. New York, NY: John Wiley & Sons, Ltd; 1999.

18. Washington C. Particle size analysis in pharmaceutics and other industries: theory and practice. West Sussex: Ellis Horwood; 1992.

19. Scarlett B. Particle Size Analysis. New York, NY: Chapman and Hall; 1994.

20. Allen T. Powder Sampling and Particle Size Determination. Amsterdam: Elsevier; 2003.

21. Merkus HG. Particle Size Measurement. Berlin: Springer; 2009.

22. Morton DAV, Staniforth JN. The challenge of the new: device-formulation matching in dry powder inhaler systems. Pharmaceutical Manufacturing and Packing Sourcer 2005; spring.

23. Edwards DA, Hanes J, Caponetti G, Hrkach J, Ben-Jebria A, Eskew M, et al. Large porous particles for pulmonary drug delivery. Science 1997;276(5320):1868–1872.

24. Trofast EA, Briggner L-E. Process for Conditioning Substances. WO patent 95/05805. 1995.

25. Dunbar CA, Hickey AJ, Holzner P. Dispersion and characterization of pharmaceutical dry powder aerosols. Kona 1998;16:7–44.

26. York P, Kompella UB, Shekunov BY. Supercritical fluid technology for drug product development. New York, NY: Taylor and Francis; 2005.

27. Zhou Q, Morton DAV. Drug–lactose binding aspects in adhesive mixtures: controlling performance in dry powder inhaler formulations by altering lactose carrier surfaces. Advanced Drug Delivery Reviews. In press.

28. Sou T, Orlando L, McIntosh MP, Kaminskas LM, Morton DAV. Investigating the interactions of amino acid components on a-mannitol-based spray-dried powder formulation for pulmonary delivery: a design of experiment approach. International Journal of Pharmaceutics 2011;421(2):220–229.

29. Allen T. Particle Size Measurement Volume 2: Surface Area and Pore Size Determination, 5 Ed. New York, NY: Chapman and Hall; 1997.

30. Webb PA, Orr C. Analytical Methods in Fine Particle Technology. Norcross, GA: Micromeritics; 1997.

31. Buckton G, Gill H. The importance of surface energetics of powders for drug delivery and the establishment of inverse gas chromatography. Advanced Drug Delivery Reviews 2007;59(14):1474–1479.

32. Das SC, Larson I, Morton DAV, Stewart P. Determination of the polar and total surface energy distributions of particulates by inverse gas chromatography. Langmuir 2011;27(2):521–523.

33. Buckton G. Characterisation of small changes in the physical properties of powders of significance for dry powder inhaler formulations. Advanced Drug Delivery Reviews 1997;26(1):17–27.

34. Buckton G, Darcy P. Assessment of disorder in crystalline powders—a review of analytical techniques and their application. International Journal of Pharmaceutics 1999;179:141–158.

35. Burnett D, Malde N, Williams D. Characterizing amorphous materials with gravimetric vapour sorption techniques. Pharmaceutical Technology Europe 2009;21(4):41–45.

36. Begat P, Morton DAV, Price R, Staniforth JN. The cohesive–adhesive balances in dry powder inhaler formulations I: direct quantification by atomic force microscope. Pharmaceutical Research 2004;21:1591–1597.

37. Zhou Q, Gegenbach T, Denman J, Das S, Qi L, Zhang HL, et al. Characterization of the surface properties of a model pharmaceutical fine powder modified with a

pharmaceutical lubricant to improve flow via a mechanical dry coating approach. Journal of Pharmaceutical Sciences. In press.

38. Shur J, Price R. Advanced microscopy techniques to assess solid-state properties of inhalation medicines. Advanced Drug Delivery Reviews. In press.

39. Schulze D. Powders and Bulk Solids Behavior, Characterization, Storage and Flow. Berlin: Springer; 2008.

40. Thalberg K, Lindholm D, Axelsson A. Comparison of different flowability tests for powders for inhalation. Powder Technology 2004;146:206–213.

41. Rhodes M. Principles of Powder Technology. Chichester: John Wiley & Sons, Ltd; 1990.

42. Kaye BH. Powder Mixing, Volume 10. New York, NY: Chapman and Hall; 1997.

43. Campbell C, Keaveny B. Key technologies and opportunities for innovation at the drug substance–drug product interface. In Houson I, editor. Process Understanding: For Scale-Up and Manufacture of Active Ingredients. Chichester: John Wiley & Sons, Ltd; 2011.

44. Freeman R. Measuring the flow properties of consolidated, conditioned and aerated powders—a comparative study using a powder rheometer and a rotational shear cell. Powder Technology 2007;174:25–33.

45. Behara SRB, Larson I, Kippax P, Morton DAV, Stewart PJ. An approach to characterising the cohesive behaviour of powders using a flow titration aerosolisation based methodology. Chemical Engineering Science 2011;66:1640–1648.

46. Tuley R, Shrimpton J, Jones MD, Price R, Palmer M, Prime D. Experimental observations of dry powder inhaler dose fluidization. International Journal of Pharmaceutics 2008;358:238–247.

47. Coates MS, Tang P, Chan HK, Fletcher K, Raper JA. Characterization of pharmaceutical aerosols for inhalation drug delivery. In Sigmund W, El-Shall H, Shah DO, Moudgil BM, editors. Particulate Systems in Nano- and Biotechnologies. New York, NY: Taylor and Francis; 2008.

48. Gumbleton M, Taylor G. Theme issue on challenges & innovations in effective pulmonary systemic & macromolecular drug delivery. Advanced Drug Delivery Reviews 2006;58/9–10.

49. Weers JG, Tarara TE, Clark AR. Design of fine particles for pulmonary drug delivery. Expert Opinion on Drug Delivery 2007;4(3):297–313.

50. Vehrig R. Pharmaceutical particle engineering via spray drying. Pharmaceutical Research 2008;25(5):999–1022.

51. Rogueda P. Novel hydrofluoroalkane suspension formulations for respiratory drug delivery. Expert Opinion in Drug Delivery 2005;2(4):625–638.

52. Lewis D. Metered-dose inhalers: actuators old and new. Current Opinion on Drug Delivery 2007;4(3):235–245.

53. Morton DAV, Jefferys D, Ziegler LR, Zanen P. Workshop on devices: regional issues surrounding regulatory requirements for nebulizers. Proceedings of Respiratory Drug Delivery Europe Conference. Boca Raton, FL: Davis Healthcare International; 2009. pp. 129–148.

54. Finlay WH. The Mechanics of Inhaled Pharmaceutical Aerosols: An Introduction. San Diego, CA: Academic Press; 2001.

55. McCallion ONM, Taylor KMG, Bridges PA, Thomas M, Taylor AJ. Jet nebulisers for pulmonary delivery. International Journal of Pharmaceutics 1996;130:1–11.

56. Taylor KMG, McCallion ONM. Ultrasonic nebulisers for pulmonary delivery. International Journal of Pharmaceutics 1997;153:93–104.

57. Traini D, Young PM, Rogueda P, Price R. In vitro investigation of drug particulates interactions and aerosol performance of pressurised metered dose inhalers. Pharmaceutical Research 2007;24(1):125–135.

58. Noymer P, Myers D, Glazer M, Fishman RS, Cassella JV. The staccato system: inhaler design characteristics for rapid treatment of CNS disorders. Respiratory Drug Delivery 2010;1:11–20.

6

Aerodynamic assessment for inhalation products: fundamentals and current pharmacopoeial methods

Francesca Buttini[1], Gaia Colombo[2], Philip Chi Lip Kwok[3], and Wong Tin Wui[4]

[1]*Department of Pharmacy, The University of Parma, Parma, Italy*
[2]*Department of Pharmaceutical Sciences, The University of Ferrara, Ferrara, Italy*
[3]*Department of Pharmacology and Pharmacy, LKS Faculty of Medicine, The University of Hong Kong, Hong Kong, China*
[4]*Faculty of Pharmacy, Universiti Teknologi MARA, Puncak Alam, Selangor, Malaysia*

6.1 Introduction

Impactors are the official instruments for determining the particle size distribution of aerosols generated from medical inhalers and nebulizers. They directly measure the aerodynamic particle size, which affects how particles move in an airstream. At the same time, impactors provide the only means of separating the mass of active pharmaceutical agents into different size ranges, as well as other non-physiologically active components of the formulation. The aerosol clouds generated by inhalers typically have a single drug or a drug combination, lactose, or propellant as their main excipient, as well as adjuncts and other additives to improve the product performance.

Methods of aerosol particle sampling and sizing are executed by fluxing the aerosol through a series of stages of know dimensions. The aerosol is carried by the airflow inside the impactor and particles impinge upon a flat collecting surface, called plate. The orifice diameter of the stage entrance and the distance from the

Inhalation Drug Delivery: Techniques and Products, First Edition. Paolo Colombo, Daniela Traini, and Francesca Buttini.

collecting flat surface are kept fixed. Particles with a high inertial force are unable to follow the streamlines and collide upon the first stages of the instrument. Smaller particles are able to remain airborne, following the streamlines through the successive impaction plates; with smaller and smaller sizes—and favorable aerodynamic properties—they impact upon lower and lower stages [1].

Impactors are used not to predict in vivo deposition but to carry out quality assessment. In vitro, the velocity of the particles increases from the first stage through successive stages; in vivo, velocity decreases from the throat to the alveoli.

Metered-dose inhalers (MDIs) consist of a solution or suspension formulation of drug in propellant and utilize a pressurized canister and valve to sample a precisely metered volume of the formulation. Dry powder inhalers (DPIs) on the other hand are made of finely divided solid drug particle clusters with or without particulate excipients. Pulmonary administration with a DPI requires an inhalation effort by the patient in order to deaggregate particle clusters. A higher inhalation flow rate produces a greater level of powder deaggregation, giving rise to individual particles for effective administration. In this case, the deliverable and respirable doses are dependent on the inhalation flow rate and tend to be inconsistent with each administration. This difficulty is further complicated by varying the aggregation degree of the powder in the inhaler during storage, under the influence of relative humidity. Recent DPIs exploit compressed air or battery-powered impellers to deaggregate the powder clusters without relying on patient inhalation flow rate.

In vitro aerodynamic assessment of the size and size distribution of particles, the fine particle fraction (FPF), and/or the mass that can reach the lungs provides a measure of inhaler consistency in drug delivery and in drug deposition pattern in the lungs [2]. Particles for inhalation have to be designed by tailoring their size and physicochemical properties. Their functionality is determined by aerodynamic assessment of the aerosol.

The currently accepted standardized test methods and apparatuses for inhaler quality assessment are documented in pharmacopoeias such as the *European Pharmacopoeia* and the *United States Pharmacopoeia* [3,4]. Broadly, the aerodynamic profile of an aerosol cloud from an inhaler or nebulizer can be examined by means of a twin-stage impinger, Andersen Cascade Impactor (ACI), Marple–Miller impactor, Multi Stage Liquid Impinger (MSLI), or Next Generation Impactor (NGI).

6.2 Impactor/impinger design

Impactors are apparatus in which particles impact on a dry surface (i.e. a collecting plate). In impingers, particles deposit instead on a liquid surface. The general principle of inertial impaction is applied to both impactors and impingers.

Inertia impaction has been used to fractionate aerosolized particles in component sizes. It provides a direct measure of the aerodynamic size of particles. This is determined by collecting particles on an inertial collection apparatus, such as a multistage impactor. Dispersed in an airflow, particles will follow the direction of airflow until they lose inertia as a result of friction between particles and the surrounding medium. Then the particles are relaxed and moved into a different direction of flow. The "relaxation time" is the time required for a particle to adjust or relax its velocity to a new condition of drag forces. In an aerosol, small particles relax rapidly. They possess a lower level of momentum (the motion of a body, equal to the product of its mass and velocity). Small particles thus do not tend to be collected, while large particles impact on collection surfaces placed in the path of the original direction of airflow. Consequently, aerosolized particles are classified according to their differences in size.

Both impactors and impingers are built based on the concept of inertia impaction. The ACI and NGI (Figure 6.1) are the main instruments used in the development and quality characterization of inhalation products. They share many similar instrumental elements in their design and test protocol with the Marple–Miller impactor and MSLI. Typically, both impactors and impingers are a sequential assembly of an inhaler mouthpiece, induction port, series of collection stages, filter system, flow-control valve with absolute-pressure transducers, two-way solenoid valve, timer, and vacuum pump. Sometimes additional components are required, such as a preseparator, which is added to remove particles larger than 20 μm in the aerodynamic assessment of a DPI. The preseparator, placed between the induction port (throat) and the first impactor stage, is particularly useful in capturing the large particles of the lactose carrier when a dry powder blend formulation is being tested.

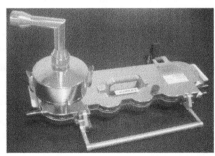

Figure 6.1 Cascade impactors, from left to right: multistage liquid impinger (MSLI), next-generation impactor (NGI), Andersen cascade impactor (ACI). Courtesy of Copley Scientific Limited, UK

The main components of an impactor for the aerodynamic measurement of inhalation product are the induction port, impactor stages, and vacuum pump. The induction port (an aluminum or stainless-steel tube with a 90° bend) serves to simulate the mouth and throat of a patient, and to direct the airflow to align with the positions of connected impactor stages. The vacuum pump creates an airflow through the impactor stages, reproducing at the induction port the peak flow value of an inspiration act. The flow is kept constant for the duration of the test.

The impactor or impinger is made of a series of four to eight tightly sealed stages. Each impactor stage is characterized by a nozzle or a set of nozzles and an impaction plate [5]. Like the ACI, the most recently developed impactor (the NGI) consists of eight stages. The aerosol particles are accelerated through nozzles to an impaction plate. The larger particles will impact on to the plate and be collected, while smaller particles will travel with the airflow to impact on a later plate. The impactor stages collect aerosolized particles into various fractions of different particle sizes. The final integral part of an impactor or impinger is a filter or microorifice collector (MOC), which collects the nonimpacted particles from the airflow. Particles <0.1 μm deposit by Brownian diffusion: an irregular wiggling motion in still air from a region of higher concentration to a region of lower concentration. Brownian diffusion, interception, charge, sedimentation, and inertia impaction are all deposition mechanisms which contribute together to the deposition of particles.

6.3   Aerodynamic assessment

Aerodynamic assessment is performed through a series of experiments and successive calculations aimed at determining the aerodynamic size distribution of aerosol particles. The aerodynamic diameter is the characteristic parameter used to express a particle's ability to follow an airstream. It assesses in vitro the respirability of a product. The aerodynamic assessment is performed using impactors, by measuring the number of particles deposited on the various stages. An impactor stage has a "cut-off size": all particles above a certain size are collected, while all particles below that size pass through. This sequential separation has the effect of classifying the particles into groups according to their aerodynamic diameter. From the amount of sample collected at each stage, the fraction of the total mass in each aerodynamic size range can be determined using the procedure described in Section 6.8.

The aerodynamic diameter is a characteristic equivalent spherical diameter derived using the Stokes equivalent diameter.

- The Stokes diameter (d_{St}) is the diameter of a sphere with the same density and settling velocity as the particle.

- The aerodynamic diameter (d_{ae}) is the diameter of a sphere of unit density that has the same settling velocity as the particle.

Stokes' law describes the particle sedimentation under gravity in a fluid. The application of Stokes' law allows the determination of the settling velocity of aerosol particles undergoing gravitational settling in air. The terminal settling velocity, V_{ts}, of a particle increases rapidly with particle size, being proportional to the square of the particle diameter. V_{ts} is calculated by the formula:

$$V_{ts} = \frac{\rho \, d_{st}^2 g}{18\eta} \tag{6.1}$$

where ρ is the density of spherical particles, g is the acceleration due to gravity, and η is the viscosity of the fluid where the particle settles. Hence the relation which links the two diameters is as follows:

$$d_{ae} = d_{st}\sqrt{\frac{\rho}{\rho_0}} \tag{6.2}$$

Therefore, the aerodynamic diameter is the diameter of a hypothetical spherical particle of unit density (i.e. $\rho_0 = 1.00\,\text{g/cm}^3$) that settles in air at the same falling velocity as the real particle. A correction factor, called the dynamic shape factor (χ) (included in d_{St}), is introduced into the relationship between d_{ae} and d_v (equivalent volume diameter) in order to consider the effect of shape on particle motion. The shape factor is equal to 1 for a sphere and—except for certain streamlined shapes—the dynamic factor is greater than 1.0. The shape factor introduction allows the aerodynamic diameter of an aerosol to be calculated using the equivalent volume diameter by the formula:

$$d_{ae} = d_v\sqrt{\frac{\rho}{\rho_0 \chi}} \tag{6.3}$$

where d_v is the spherical equivalent volume diameter and χ is the dynamic shape factor. Owing to deformation by airflow stresses, the droplets of MDI product are not completely spherical. Inferring from this equation, the nonspherical particles are likely to exhibit a smaller aerodynamic diameter.

Principally, the aerodynamic diameter of an aerosol is dependent on particle properties such as geometric size, shape, density, and surface morphology. Porous particles will have a smaller aerodynamic diameter than dense particles. The aerodynamic property of liquid aerosol experiences a greater propensity of complication than that of DPI products, since solvent evaporation, phase separation, and the influences of additives greatly modify the properties of aerosol particles.

The particle size distribution of an aerosol typically follows a log-normal distribution pattern. It can be characterized by mass median size, which denotes diameter where 50% of the aerosolized particles are larger or smaller than the stated size. The distribution of particle size is described by geometric standard deviation (GSD). Using an impactor or impinger, which separates particles with an airflow and quantifies drug mass against their sizes, the size of an aerosol is termed the "mass median aerodynamic diameter" (MMAD). The computation of the MMAD of an aerosol proceeds by plotting on a log-probability scale the cumulative percentage of drug mass retained on each stage with the smaller cut-off diameter versus the logarithm of the cut-off diameter of the stage. The linearity of the relationship between the cumulative percentage of drug mass and log cut-off diameter is determined by least-squares regression. MMAD is the particle size below which 50% of the population falls and is often denoted as d_{ae50}. The GSD of aerosol size distribution, which expresses the particle distribution around the MMAD value, can be calculated from the slope of the same plot, as illustrated in Section 6.8 [4].

The aerodynamic diameter depends on the airflow around the particle, which is expressed through the Reynolds number (Re):

$$Re = \frac{\rho V d}{\eta} \qquad (6.4)$$

where ρ is the density of the fluid flow, V is the relative velocity between the fluid and an object such as a particle, d is the diameter of the particle, and η is the viscosity of the fluid in which particle settles.

Care should be taken to note that the density is that of the gas, not the aerosol particle—a frequent source of confusion in the application of Re to aerosol particles.

Re is a dimensionless number that indicates whether the flow of a fluid (liquid or gas) is absolutely steady (laminar flow) or steady on average but with small unsteady changes (turbulent flow).

The human respiratory system between the trachea and the terminal bronchioles exhibits a laminar airflow with an Re of 0.01–2. The airflow through the impactor has an Re of 0.1–20. The Stokes flow regime (laminar flow) depicts an aerodynamic diameter related to particle deposition by sedimentation and Brownian diffusion. The use of Stokes flow in the aerodynamic assessment of aerosol at high airflow rates is envisaged to lead to systematic errors. At $Re > 0.1$, inertial impaction represents the major deposition mechanism at upper airways.

The performance of a product for inhalation depends on both the aerodynamic and the geometric particle size distribution of the aerosol. The geometric particle size and size distribution determine the interparticle interactions, delivered dose and dose uniformity, drug dissolution rate, and cellular particle uptake. The aerodynamic particle size distribution and MMAD of aerosolized particles govern the respirable dose and particle deposition pattern in the lungs. Aerosol particles are deposited in the lungs by three main mechanisms: impaction, sedimentation, and Brownian diffusion. Particles with an MMAD ranging between 1 and 5 μm are known to deposit in bronchial and alveolar regions and demonstrate pulmonary penetration. Smaller particles within this size range exhibit a greater propensity of lung penetration. They reach distal airways and have more peripheral deposition. They are suitable for delivery of anti-inflammatory drugs and drugs with systemic effects which have to be absorbed in the alveolar region, such as peptides and proteins. Nonetheless, larger particles are more efficacious where the treatment of asthma is concerned, since the deposition in this case is required in the upper airways. The depositional target of particles of a size of 5 μm is the upper lung airway, where they give rise to a direct bronchodilation effect.

6.4 Inertial impaction and cut-off diameter

Particle Stokes number is a dimensionless parameter which governs the efficiency of particle impact on a collection plate, defined for an impactor as the ratio of particle stopping distance at the average nozzle exit velocity (U) to the nozzle radius. It is calculated according to the following expression:

$$Stk = \frac{U_0 \, \rho \, d^2 \, C_c}{9 \, \eta \, D_n} \qquad (6.5)$$

where U_0 is the average fluid velocity to the nozzle, ρ is the particle density, d is the particle diameter, C_c is the Cunningham slip correction factor, η is the dynamic viscosity of air, and D_n is the nozzle diameter.

The Cunningam correction factor is a factor applied to Stokes' law for particles less than 1 μm in diameter which settle faster than predicted by Stokes' law because there is a "slip" at their surface.

The equation assumes that all particles are spherical, that their Re is much less than 1 (<0.1), and that their density is greater than air density. A particle will impact on to a plate if its Stokes number is larger than approximately unity, which translates to a need for a longer relaxation time.

The collection efficiency is the fraction of particles passing through impactor nozzles which are retained on impaction plates. All particles larger than an impactor stage's cut-off diameter will impact on that stage and all particles smaller than the cut-off diameter will follow the airflow through to successive impactor stages. This cuts the distribution of mass into a series of size ranges called "dimensional classes."

The cut-off diameters of impactor stages must be regularly calibrated, as nozzles and collection plates can corrode with time and usage [6].

In testing a DPI using an impactor or impinger, an airflow rate between 28 and 100 L/min may be used, since DPIs have dissimilar powder flow resistances and dispersion efficiencies [7]. Using a different airflow rate Q from Q_n (60 L/min) in association with the previously calibrated cut-off diameter, the new cut-off diameter of each stage at Q can be calculated using a preestablished equation:

$$D_{50Q} = D_{50Qn}\left(\frac{Q_n}{Q}\right)^{\frac{1}{2}} \qquad (6.6)$$

where $D_{50}Q$ is the cut-off diameter at airflow rate Q (60 L/min) and n is the nominal cut-off value determined by Q_n.

In the case of NGIs, a similar equation may be used to compute the cut-off diameter of impactor stages subjected to different airflow rates, except that the exponent value (x) varies with the impactor stage:

$$D_{50Q} = D_{50Qn}\left(\frac{Q_n}{Q}\right)^{n} \qquad (6.7)$$

Given an airflow rate Q higher than Q_n, the cut-off diameter of an impactor stage becomes smaller. The drug mass distributed across impactor stages is expected to exhibit a smaller aerodynamic size. In the calibration of impactor stages and testing of an inhalation product, the quantum of airflow rate will be specified. Each impactor stage must be calibrated and/or its cut-off diameter characterized at the airflow rate employed in aerodynamic testing of the inhalation product.

6.5 Pharmacopoeial procedure

The aerodynamic assessment of inhalation product begins with the assembly of the impactor or impinger. The volumetric flow rate at the entrance to the induction port must be accurately set before every determination of aerodynamic particle size distribution. The airflow rate used in testing MDIs is kept at 28–30 L/min, but it can vary between 28 and 100 L/min in the testing of DPIs. For the vast majority of DPIs, it is unnecessary to coordinate breathing with the activation, since the patient simply inhales deeply to access the drug. It follows that the efficacy of the administration is dependent on the strength and duration of the patient's inspiration. Furthermore, different inhalers provide varying degrees of resistance to inhalation. The in vitro assessment approximates the in vivo inspiration conditions, and the strength and the duration of the patient's inspiration in particular are simulated by, respectively, the flow rate used and the time it takes the flow to pass through the device. In DPIs, the *United States Pharmacopoeia* states that the airflow rate corresponds to a pressure drop of 4 kPa over the device, and a duration consistent with the withdrawal of 4 L of air from the mouthpiece of inhaler should be employed. These parameters are described in the section "General Control Equipment" in the *United States Pharmacopoeia* [4] and the section "Experimental Set Up" in the *European Pharmacopeia* [3,8].

6.5.1 The duration of in vivo inspiration needed for 4 L

The volume of air drawn through an inhaler during testing is kept at 4 L, as this volume represents the normal inspiratory capacity of an average-sized adult male weighing approximately 70 kg. A pressure drop of 4 kPa over the DPI broadly represents the pressure drop generated by patients using the device [9].

Prior to testing of a DPI, the intended airflow rate is set. The duration is simulated through the use of a two-way switching valve connected to a vacuum pump. The operation of the solenoid valve and hence the duration of the cycle is controlled by means of a timer. The valve is connected to both the impactor and the vacuum pump. In pre-test mode, the solenoid valve is in the closed position, such that no flow passes through the test apparatus.

On initiation of the test, the solenoid valve switches, so that flow now passes through the test apparatus and hence the inhaler under test. On expiration of the preset time, the solenoid closes again and the "inhalation" cycle is complete.

By using the timer to control how long the solenoid valve is open for, it is possible to control the volume of air drawn through the inhaler to the 4 L specified. It is also possible to establish for how many seconds (T) the valve must be in open mode in order to obtain 4 L of air ($\pm5\%$) at the test flow rate Q_{OUT}:

$$T = 240/Q_{OUT} \tag{6.8}$$

For example, if the flow rate Q is 100 L/min, the timer should be set to 2.4 seconds. If the flow rate is 60 L/min, the timer should be set for 4 seconds. And if the flow rate is 30 L/min, the timer should be set for 8 seconds. The inspiration time is important for the DPls.

Q depends on the characteristics of the device used.

6.5.2 Setting flow rate, Q, to give a pressure drop of 4 kPa

To set the correct flow rate to be used in the test, it is first necessary to establish the flow rate that produces a pressure drop comparable with that found in vivo when using the particular inhaler under study. The flow rate to be used is the flow rate that produces a pressure drop through the inhaler compatible with the resistance of the device. Both the *European* and the *United States Pharmacopoeia* suggest a pressure drop over the inhaler of 4 kPa is broadly representative of the pressure drop generated by patients using powder inhalers during inhalation [3,4]. The pressure drop created by the air drawn through an inhaler can be measured directly by comparing the absolute pressure downstream of the inhaler mouthpiece to the atmospheric pressure.

Using a flow-control valve, it is now possible to adjust the flow rate from the vacuum pump to produce the established pressure drop of 4 kPa. By replacing the inhaler with a flow meter, it is then possible to measure the flow rate required to produce this pressure drop. This flow rate should be used for determination of both the delivered dose and the particle size. The only exception to this criterion is that if the flow rate required to produce a 4 kPa pressure drop is greater than 100 L/min, 100 L/min should be used.

6.5.3 Flow-rate stability

Having set the required parameters to control the peak and the duration of the simulated inspiration, there is one final variable which needs to be considered: the flow-rate stability. The flow rate can be assumed to be stable if the ratio P_3/P_2 is ≤ 0.5, where P_2 represents the pressure at a point nearer the impactor or impinger and P_3 represents the pressure at a point nearer the vacuum pump. This parameter is important to ensuring that the flow rate achieved through the inhaler is unaffected by pump fluctuations and minor changes in flow resistance upstream of the flow-control valve.

6.6 Cascade impactor: general set-up and operation

Proper set-up and operation of an impactor is critical for correct particle size measurement. A test failure can arise if the system is not properly assembled or the airflow is not checked prior to running the analysis. Errors can come from the collection plates or filter being wrongly located in the ACI, from air leakage into the apparatus, or from an improper alignment between inhaler mouthpiece and induction port [10]. During testing of the inhalation product, the inhaler is first primed or loaded and inserted in axis with the entry to the induction port (Figure 6.2). Each device has to fit properly with the induction port, through the use of a house-made rubber adaptor.

The vacuum pump is then run at the required airflow rate and the inhaler is left in place for 20 seconds to enable complete release of the aerosol prior to switching off the solenoid valve. Repeat the loading and discharge procedure when additional

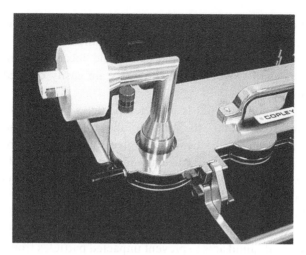

Figure 6.2 Alignment between the RS01 DPI inhaler (Plastiape, IT), mouthpiece, and induction port of the NGI. RS01 DPI - Plastiape. Courtesy of Plastiape IT, Courtesy of Copley Scientific Limited, UK

doses are required. The inhaler mouthpiece, induction port, impactor stages, and filter system are then washed with a drug solvent. The choice of solvent for a particular drug formulation is critical since it affects dissolution of the drug and thus recovery from the impactor collection. The solvent can also influence the quantification of drug.

The collected samples, diluted when needed, are subjected to drug mass assay using ultra-violet spectrophotometry or a high-performance liquid chromatography (HPLC) technique. The weight loss of the inhaler can similarly be determined to derive the amount of drug delivered, where a DPI formulated with a single drug without excipients is being tested.

For most formulations, multiple actuations are required to achieve the analytical sensitivity for drug-content determination. The number of doses discharged in the impactor affects both the total mass collected and the mass of drug on each stage, but the shape of the aerodynamic distribution will be unaffected since the size-fractionating capability of the impactor will not be impaired.

The aerodynamic assessment of inhalation product using an impactor or impinger is often accompanied by various complications. The collection surfaces of impactor stages for small particles tend to be overloaded with particle mass. An approach to eliminating such incidents includes the use of adhesive or fibrous layers on each impaction plate in order to reduce slippage of particles to impactor stages with lower cut-off diameters. In an impactor or impinger, the interstage drug loss should not be more than 5% of the total drug mass delivered. In the event that interstage drug loss is more than 5%, the drug mass assay may include drug lost at the wall as well as at the collection plate.

A complication that must be controlled is the air leakage through the system, arising from incorrectly located defective seals into the impactors. This can affect the size distribution, depending on the extent and location of the leakage. In particular, the standard O-rings used with ACIs are easily cracked by repeated exposure to solvents [10].

Another phenomenon to avoid is impacted particles being re-entrained in the flowing airstream. To avoid this, the surface of the collection plate at each stage of the impactor can be coated with silicone fluid, glycerol, or another substance prior to formulation testing. This effect is particularly important for dry powder formulation and can occur with some MDI products, though it may be not required for aerosols comprising liquid droplets [11,12].

Using an MSLI, silicone fluid or glycerol coating of the collection plate is not needed, as each stage must contain 20 mL liquid before testing, except the final stage. The dispensed liquid acts to prevent impacted particles from re-entraining in the airstream, as well as to simulate realistic impaction processes by particles on liquid in the lungs. Water is the preferred choice of liquid solvent in this case. However, organic solvent may be used for sparingly water-soluble drugs for ease of drug dissolution and subsequent drug mass assay. Problems such as solvent evaporation can arise, particularly at high flow rates with volatile organic solvents. The drug mass assay, as well as aerodynamic size and size distribution computation of inhaled aerosol, can be inaccurate. Solvent evaporation can be addressed by using an internal standard with a known concentration in solvent [10].

6.7 Impactor/impinger characteristics

6.7.1 Twin-stage impinger (or glass impinger, apparatus A)

The twin-stage impinger was the first apparatus adopted and its value is recognized by the *European Pharmacopoeia* [3]. This apparatus is relatively simple, easy to use and to assemble, and represents a quick method for measuring the respirable fraction of formulations.

It is manufactured only from glass, so that it is not inclined to corrosion in the same way as conventional metallic impactors. The impinger consists of two round glass flasks connected by glass tubes and a final conical flask. Samples are collected into a suitable liquid for the analysis. The apparatus is designed to operate at 60 L/min and involves a stage with a cut-off diameter of 6.4 μm [3]. Stage 1 is constituted by parts B, C, and D (Figure 6.3) of the instrument and collects particles bigger than the cut-off value. Particles smaller than 6.4 μm settle in Stage 2, particularly parts E, G, and H (Figure 6.3).

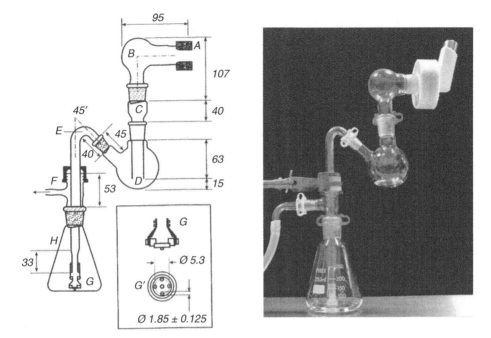

Figure 6.3 Twin-stage impinger: *European Pharmacopeia* [3] dimensional specification (kindly supplied by Copley Scientific Limited, UK) (left) and assembled apparatus (right).

The instrument operates on the principle of liquid impingement, dividing the dose emitted from the inhaler into respirable and nonrespirable portions. The nonrespirable dose is supposed to impact on the oropharynx and is subsequently swallowed—the "oropharynx" is considered the back of the glass "throat" and the upper impingement chamber (collectively described as Stage 1). Particles smaller than 6.4 μm escaping the impingement chamber (D) represent the respirable dose that penetrates the lungs (Stage 2). Prior to testing, 7 mL of a suitable solvent is typically dispensed into the upper impingement chamber (D) and 30 mL into the lower impingement chamber (G). After the test, the active drug ingredient collected in the lower impingement chamber is assayed and expressed as a respirable fraction of the delivered dose.

While this apparatus results in a relatively inefficient size-selective sampling, it has been used in the early phases of development because of its simplicity and the short time required to carry out a test.

6.7.2 Multistage liquid impinger (MSLI)

The MSLI could be considered an evolution of the twin-stage iminger. It is a five-stage cascade impactor and is described by the *European* and the *United States*

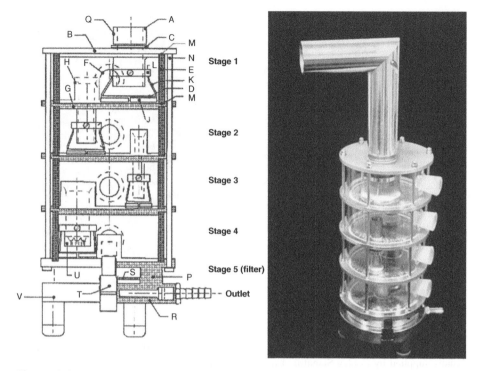

Figure 6.4   Multistage liquid impinger (MSLI): scheme (left) and assembled apparatus (right). Courtesy of Copley Scientific Limited, UK

Pharmacopoeias [3,4]. It is used to determine the aerodynamic size distributions of DPIs, MDIs, and nebulizers.

The MSLI (Figure 6.4) is available in a range of materials—aluminum, 316 stainless steel, or titanium—and is fitted with PTFE seals as standard. This choice allows flexibility in terms of corrosion resistance, weight, and cost.

The pharmacopoeia describes the same induction port as the other cascade impactors. The first four stages have glass walls and stage 5 comprises an integral paper filter, to ensure the capture of remaining particles, avoiding the damaging of the aspiration pump.

The MSLI has, by design, no interstage losses, and is suitable for use throughout the range 30–100 L/min. The design is such that at a flow rate of 60 L/min (nominal flow rate Q_n), the cut-off diameters of Stages 1, 2, 3, and 4 are 13, 6.8, 3.1, and 1.7 μm respectively [4].

When the flow rate is not 60 L/min, the new cut-off-diameter values of the individual stages must be determined according to the instructions in Section 6.4.

The collection stages of the MSLI are kept moist (usually with the recovery solvent), which helps to reduce the problem of rebound of powder, avoiding the need for plate-coating.

6.7.3 Andersen cascade impactor (ACI)

This apparatus is an eight-stage cascade impactor that has been designed to measure the particle size distribution generated by MDIs and DPIs, as described by the *European* and the *United States Pharmacopoeias* [3,4]. The apparatus allows for a detailed particle size distribution determination and can be operated at various flow rates. Several stages (Figure 6.5), with jets of well-defined size, are arranged in sequence, such that particles of progressively finer size are collected as the aerosol passes through the instrument [2].

The ACI is traditionally constructed from aluminum, but it can also be manufactured from 316 stainless steel and titanium. 316 stainless steel is the pharmaceutical industry's preferred material due to its superior corrosion resistance and durability. The stages are clamped together and sealed with silicone rubber O-rings approved by the US Food and Drugs Administration (FDA) in order to obtain leak-free interstage sealing. The final stage contains a paper filter, in order to ensure the capture of the finest particles. The completed instrument, with preseparator (normally used with DPIs), includes an induction port for simulation of the throat.

The ACI, like other cascade impactors, is designed such that, as the aerosol stream passes through each stage, particles with sufficient inertia will impact upon the stage collection plate, while smaller particles, with insufficient inertia, will follow the streamlines and pass to the next impaction stage.

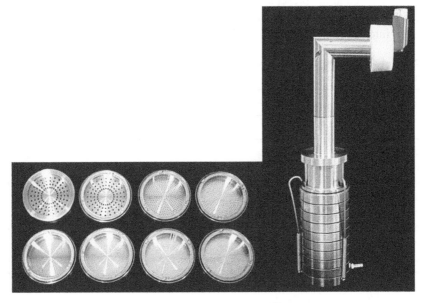

Figure 6.5 Andersen cascade impactor (ACI): the eight stages (left) and the assembled apparatus, with pre-separator (right). Courtesy of Copley Scientific Limited, UK

The standard ACI is designed for use at 28.3 L/min, but in many cases (particularly with low-resistance DPIs) it is necessary to operate at rates above this if a pressure drop across the inhaler of 4 kPa is to be achieved. However, it is important to consider that the flow rate will affect the cut-off points for each stage. An empirical equation can be used to calculate the cut-off-point changes over the range 28.3–100 L/min. The cut-off diameter for each stage follows Stokes' law—that is, the cut-off diameter is inversely proportional to the square root of the airflow—but a reduced discrimination between the cut-off points will occur as the flow rate is increased. To overcome this problem, two modified configurations are available, for operation at flow rates in the regions of 60 and of 90 L/min. These are described in the *United States Pharmacopeia* [4].

In the 60 L/min impactor version, Stages 0 and 7 are removed and replaced with two additional stages, −0 and −1. Similarly, in the 90 L/min impactor version, Stages 0, 6, and 7 are removed and replaced with Stages −0, −1, and −2. Table 6.1 shows the cut-off values when the configuration is modified and the ACI is used at 28.3 L/min [13,14].

A standard impactor can be used to test MDIs without further modification. Sometimes, a preseparator, interposed between the induction port and Stage 0 of the impactor, can be used to determine the particle size of a DPI, in order to collect the large mass of noninhalable particles, such as those of the carrier. The omission of the preseparator when it is required will have a large impact on the aerodynamic distribution, while leaving mass balance unaffected, since the mass of API that might be retained in the preseparator will reach the lower stages of the impactor. In the case of DPIs, a number the additional factors described in Section 6.5 must also be taken into account [3].

Table 6.1 Cut-off aerodynamic diameter for stages of the Andersen cascade impactor (ACI) when used at 28.3, 60, and 90 L/min

Cut-off aerodynamic diameter for stages of the ACI			
	28.3 L/min	60 L/min	90 L/min
Stage −2	–	–	>9.0 μm
Stage −1	–	>9.0 μm	5.8–9.0 μm
Stage −0	–	5.8–9.0 μm	4.7–5.8 μm
Stage 0	9.0–10.0 μm	–	–
Stage 1	5.8–9.0 μm	4.7–5.8 μm	3.3–4.7 μm
Stage 2	4.7–5.8 μm	3.3–4.7 μm	2.1–3.3 μm
Stage 3	3.3–4.7 μm	2.1–3.3 μm	1.1–2.1 μm
Stage 4	2.1–3.3 μm	1.1–2.1 μm	0.7–1.1 μm
Stage 5	1.1–2.1 μm	0.7–1.1 μm	0.4–0.7 μm
Stage 6	0.7–1.1 μm	0.4–0.7 μm	–
Stage 7	0.4–0.7 μm	–	–
Filter	<0.4 μm	<0.4 μm	<0.4 μm

6.7.4 Next-generation impactor (NGI)

The NGI was developed by an industrial consortium, specifically for particle-size-distribution-profile determination of pharmaceutical aerosols, to resolve dissatisfaction over several aspects of the ACI's performance.

The use of the NGI for inhaler testing first appeared in the *European Pharmacopoeia* [3] in Section 2.9.18 of Supplement 5.1 of the 5th Edition, September 16, 2004. A similar description first appeared in the *United States Pharmacopeia* [4] Volume 28, Section <601>, January 1–March 31, 2005. It is the last impactor and the only horizontal described in either *Pharmacopeia*. It was originally developed to facilitate the recovery procedure without affecting the size distribution of the active pharmaceutical ingredient (API).

The NGI (Figure 6.6) consists of a lid, seal body, nozzle pieces, collection cups, cup tray, and bottom frame. Most importantly, there are eight nozzle pieces in the NGI, corresponding to seven size-fractionation stages, and an MOC, which takes the place of a final filter. Major accessories include the NGI induction port, preseparator, external final filter, and internal final filter [15].

The *US* and the *European Pharmacopoeia* prescribes the use of a preseparator only for dry powder with large solid-carrier testing. Its main function is to retain the large, inert particles according to the specifications of the ACI.

The routine assembly of an NGI is schematically explained as follows: Place clean and dry cups into the slots on the cup tray and pick up the tray by the handles. Align the tray so that the cups are over the corresponding openings in the bottom frame. Lower the cup tray into place. Swing the impactor lid, with the attached seal body, into place on top of the collection cups. Lower the handle to see all the components of the impactor together. Use the NGI preseparator only if your procedure calls for the use of a preseparator. In case the preseparator is to be used, place the male-taper end of the preseparator into the inlet of the impactor. Place the male taper of the NGI induction port into the inlet of the NGI or of the preseparator. Attach a mouthpiece adapter that fits with your inhaler to the NGI induction port; this adapter is often custom-made so as to ensure a leak-free connection between the inhaler and the induction port. Attach a vacuum source to the exhaust port of the impactor. Material may accumulate on either the upstream or the downstream side of the nozzles. Typically, if any deposits are found they are on the downstream or so-called "backside" of the nozzles. The material found on the backside of a nozzle piece should be less than 5% of the mass found in the collection cup underneath the nozzle piece. In the development of the NGI, the interstage passageways were found to have less than 1% of the mass of drug product captured in the collection cups. Therefore, the interstage passageways will need to be cleaned only infrequently (approximately every 20–50 determinations).

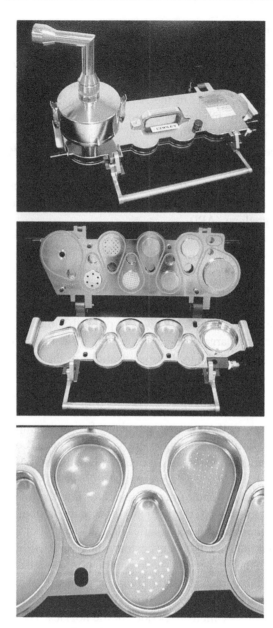

Figure 6.6 Closed NGI with USP/EP pre-separator and induction port (top). NGI in the "open" configuration (middle). NGI collection plates with powder formulation deposited inside (bottom). Courtesy of Copley Scientific Limited, UK

Table 6.2 Stage cut-off size of the NGI at 30, 60, and 100 L/min. Impactor inlet flow rate is calculated from the archival NGI calibration

Cut-off aerodynamic diameter for stages of the NGI

	x	30 L/min	60 L/min	100 L/min
Stage 1	0.54	>11.72 μm	>8.06 μm	>6.12 μm
Stage 2	0.52	6.40–11.72 μm	4.46–8.06 μm	3.42–6.12 μm
Stage3	0.50	3.99–6.40 μm	2.82–4.46 μm	2.18–3.42 μm
Stage4	0.47	2.30–3.99 μm	1.66–2.82 μm	1.31–2.18 μm
Stage5	0.53	1.36–2.30 μm	0.94–1.66 μm	0.72–1.31 μm
Stage6	0.60	0.83–1.36 μm	0.55–0.94 μm	0.40–0.72 μm
Stage 7	0.67	0.54–0.83 μm	0.34–0.55 μm	0.24–0.40 μm
MOC	-	<0.54 μm	<0.34 μm	<0.24 μm

The NGI has been calibrated with monodisperse particles at 15, 30, 60, and 100 L/min. In the range 30–100 L/min, the D_{50} values for each stage follow impactor theory rather well and are given by the equation [16,17]:

$$D_{50Q} = D_{50Qn}\left(\frac{Q_n}{Q}\right)^x \tag{6.9}$$

where $D_{50}Q$ is the cut-off diameter at the flow rate Q employed in the test and the subscript n refers to the nominal value at which Q_n is equal to 60 L/min. The values for the exponent x change with each stage calculation and are listed in Table 6.2.

6.8 Data analysis

Section 601 of the *United States Pharmacopeia* [4] sets out the most common way to analyze size distribution data: namely, by plotting the cumulative percentage of the drug mass recovered in stages smaller than a stated aerodynamic diameter.

An aerosol commonly contains particles of various sizes. Aerodynamic size distribution data from impactors are obtained as mass fractions of APIs collected in each part of the system. Figure 6.7 shows the API deposition location in both nonsizing (induction port, preseparator) and size-fractioning components. The deposition profile is typical for each formulation and is a useful tool for compaing products for inhalation; for example, a reference versus a generic test product [18].

By assaying the amount of drug deposited on the various stages, it is possible to calculate the fine particle dose (FPD) and FPF, and further the MMAD and GSD, of the active drug particles collected.

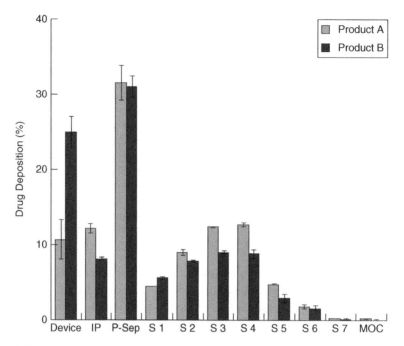

Figure 6.7   Drug mass collected (%) on each stage of a next-generation impactor (NGI) for two different formulations tested. IP = induction port, P-Sep = pre-separator, S = stage, MOC = microorifice collector

Data interpretation is done by plotting the cumulative percentage undersize sampled at each stage of the impactor on a probability scale against the log of the cut-off diameter of the stages, resulting in a collection-efficiency curve (Figure 6.8).

The experimental MMAD and the GSD can be calculated from the log-probability-scale plot. The MMAD corresponds to the diameter of the particles deposited in the impactor for which 50% w/w have a lower diameter and 50% w/w have a higher diameter.

The GSD can be calculated by:

$$GSD = \sqrt{\frac{d_{84.13}}{d_{15.87}}} \tag{6.10}$$

where $d_{84.13}$ is the particle diameter (μm) when the cumulative mass (%) is equal to 84.13% and $d_{15.87}$ is the particle diameter (μm) when the cumulative mass (%) is equal to 15.87%.

FPD denotes the mass, in the milligram–microgram range, while FPF denotes the percentage relative to the total quantity of drug collected in the impactor or impinger

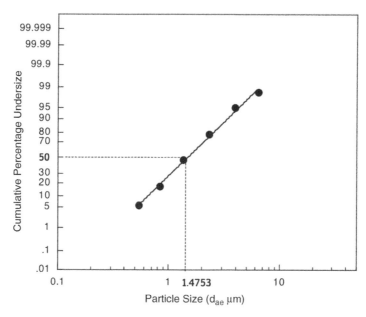

Figure 6.8 Cumulative percentage or mass below the stated aerodynamic diameter (probability scale) versus aerodynamic diameter (log scale)

that has a size equal to or less than 5 μm. The delivered dose may be discovered by weighing the inhalation formulation when it is made of pure drug only. When the formulation comprises drug plus excipients, the weight difference of an inhaler before and after drug delivery represents the amount of product delivered.

Recovery FPD and FPF can be assessed in addition to aerodynamic size. "Recovery" refers to the percentage of drug recovered from the impactor or impinger in relation to the mass of particles emitted from the inhaler. The *European Pharmacopeia* [3] establishes that the values have to be between 75 and 125% in order for the test to be considered valid.

6.9 Cleaning instructions for impactors

The accumulation of deposits in stage nozzles can affect the aerodynamic size distribution of successive determinations. If the impactor is not cleaned periodically (the frequency depends on the formulation), both mass balance and drug distribution may be affected [19]. If sample recovery requires the use of HCl or other strong acids, or the use of NaOH or other strong bases, the parts must be washed and dried immediately following sample recovery.

6.10   Test limitations

The aerodynamic measurement of DPI is conducted using a single airflow rate. However, the dispersion of dry powder and its deposition profile in the lungs is the result of inconsistent inspiratory rates.

Real-world test procedure using an impactor or impinger requires clarification. At an atmospheric relative humidity greater than 75%, the particles adhere on to the collection plate easily [20]. The cut-off diameter is nearly double across the whole range of humidity. The sensitivity of test outcome to relative humidity can be product-dependent. This problem should be addressed by practical investigation, though it is suggested that it can be resolved by a test conducted at a relative humidity lower than 70%.

The impactor technique is considered invasive, laborious, and time-consuming. In routine analysis, a simplified twin-stage impinger can be used, due to its ease of handling. The twin-stage impinger was the first device to be used for aerosol sampling and particle characterization. It provides data on the fraction of a delivered dose that is likely to be deposited in the lungs. The particles that pass through the lower part of the device are considered to be respirable and have an aerodynamic diameter less than 6.4 μm. A twin-stage impinger requires solvent in both chambers in order to collect particles. Its value compared to a cascade impactor has been debated over the years. A twin-stage impinger does not provide the total size-distribution profile of an aerosol, unlike MSLI techniques, but its use in development is very practical.

6.11   Future considerations

Nanoparticles of poorly water-soluble drugs have a high dissolution rate. Ultrafine nanoparticles with a size less than 150 nm show delayed lung clearance, increased interaction with specific proteins, and enhanced translocation from epithelium to circulation and target organs. Can the aerodynamic properties of a functional nanosize aerosol fraction be assessed with accuracy in the existing pharmacopoeial impactor or impinger set-up? Fractions less than 0.5 μm are likely to be exhaled. Is this scenario reflected in the current mode of aerodynamic assessment?

In DPI products, a small number of fine lactose particles, with sizes below 10 μm, are used as carriers to promote deaggregation of drug powder. However, aerosol performance does not depend on the inherent size of the carrier. Instead, the fine lactose particles adhered on to the large carrier particles are responsible for reducing drug–drug contact, thereby facilitating drug dispersion and deposition. Is there a need to address the aerodynamic particle size of the carrier? Specifically, does the weight percentage of fine lactose carrier particles of sizes below 10 μm need to be characterized?

6.11.1 Electrical low-pressure impactor (ELPI)

Impactor design

The electrical low-pressure impactor (ELPI) consists of a 13-stage Berner-type multijet low-pressure impactor with a diode-type corona charger situated directly above the top stage [21] (Figure 6.9). It sorts particles aerodynamically into various size fractions by impaction, much like other cascade impactors such as the ACI or NGI. However, the size range covered by conventional impactors is usually 0.3–10 μm [22], while that of the ELPI extends down to about 30 nm, because its impactor stages operate below atmospheric pressure [22] (see Figure 6.9).

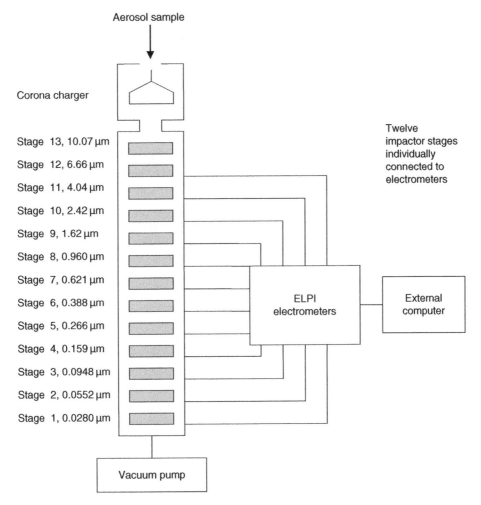

Figure 6.9 Schematic diagram of the ELPI in its original configuration. Diagram not drawn to scale

Original use of the ELPI

The ELPI was designed for near-real-time sizing of airborne particles by electrical detection [21,22]. In its original mode of operation, the corona charger is charged to +5 kV [3] and the resulting high electric field imparts a known, size-dependent level of charge to the particles introduced into the impactor by vacuum suction. The jet plate of Stage 1 acts as a critical orifice for establishing the flow rate [22], which is usually 30 L/min. All the impactor stages are electrically insulated from one another, with the last 12 stages connected individually to individual electrometers with sensitivity at femtoampere (10^{-15} A) levels. Particles subsequently deposit on to the stages according to their aerodynamic sizes and their charges are measured by the respective electrometers. Particle size distribution is then derived from the electric-current measurements according to the corona charger efficiency [21]. The computer data-collection software has an algorithm to correct for fine particle losses due to space-charge forces. Further details on the theoretical background and electronics of the ELPI are given by Keskinen et al. [21].

Particle sizing via corona charging and electrical derivation using the ELPI is an indirect method but is fast and convenient because no gravimetric or chemical analyses are required [24]. These additional assays may still be conducted afterwards if desired. The ELPI has mostly been applied in measurements or real-time monitoring of industrial and environmental aerosols. These include particulate emissions from diesel engines [25–30], vehicles [31], burning incense [32], smoke particles attached to radon decay products [33], and general atmospheric airborne particles [34–36]. Aerosols generated from MDIs [37] and DPIs [38] have also been sized by ELPI electrical detection, but in general there are relatively few pharmaceutical applications.

Modified use of the ELPI

In all the aforementioned studies, aerosol samples were charged with the corona charger. The primary interest was to derive size distributions from electrical signals. However, inherent charges on aerosol particles may also be measured by the ELPI without corona charging. Solid particles acquire charges from physical contacts between one another and between particles and inhaler components during dispersion [39]. On the other hand, disruption of the electrical double layer in liquid surfaces during atomization generates spontaneously charged droplets [39].

Although the ELPI is not a pharmacopoeial impactor, it has recently gained popularity for use in pharmaceutical particle size and charge measurements. Palonen and Murtomaa measured DPI aerosol charges with the corona charger in place but switched off [40]. However, no drug mass was determined. Glover and Chan found no significant difference in the charge measurement of MDI aerosols when the corona charger was in place but switched off versus when the charger was removed [41]. The charger was thus unnecessary and was removed to eliminate the possibility of artifacts in charge and mass measurements due to unwanted contact charging with, and deposition of particles on, the charger block [41]. A *US Pharmacopoeia* Induction Port (simulating the throat) was also connected to the ELPI inlet.

The net charge contributed by particles on each stage is derived from the electrometer readings collected from the individual stages. The deposited drug on the stages may be assayed chemically using HPLC to obtain the particle size distribution by mass [41]. The charge and mass data may be combined to yield charge-to-mass ratios, or specific charges, for the various size fractions. The advantage of using the modified ELPI to measure particle charges is the high resolution of size and charge classifications. Variations in both the mass and the net charge within the different size fractions can be discerned. This is essential for a detailed understanding on the electrostatic properties of aerosols.

The modified ELPI has been employed to measure charges generated from nebulizers, MDIs, and DPIs. Water droplets nebulized from a Sidestream jet nebulizer are positively charged [42]. Aerosols from commercial MDI products may be bipolar or unipolar in charge, depending on formulation constituents (drug, propellant, and excipients) and valve component materials [43,44]. Charge levels also vary across the different size fractions. The fundamental electrostatic properties of custom-made, simple HFA formulations have been examined with the modified

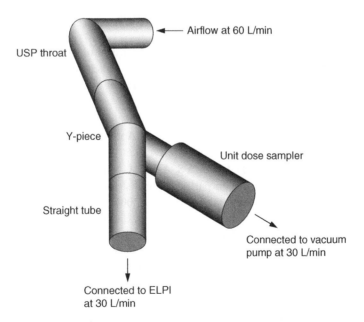

Figure 6.10 Three-dimensional representation of the modified electrical low-pressure impactor (ELPI) set-up, suitable for DPI dispersion at 60 L/min. Diagram not drawn to scale

ELPI [45,46]. As for MDIs, the charges of DPI aerosols are also influenced by the formulation and inhaler materials [47,48]. In addition, the storage and ambient relative humidity play a role in the charge profiles [48,49]. However, the ELPI only operates at a fixed airflow rate of 30 L/min, which is too low to disperse DPI aerosols. To achieve a dispersion flow rate of 60 L/min, a Y-tube and other extra parts are added to the set-up (Figure 6.10) [48]. Telko et al. [47] employed a similar set-up, but used a custom-made flow splitter with Tygon tubings and an ACI in place of the Y-tube and unit dose sampler, respectively.

References

1. Mitchell J, Newman S, Chan H-K. In vitro and in vivo aspects of cascade impactor tests and inhaler performance: a review. AAPS PharmSciTech 2007;8(4):237–248.
2. Menzies D, Nair A, Fardon T, Barnes M, Burns P, Lipworth B. An in vivo and in vitro comparison of inhaled steroid delivery via a novel vortex actuator and a conventional valved holding chamber. Annals of Allergy, Asthma and Immunology 2007;98(5):471–479.
3. *European Pharmacopoeia* 7.1. Section 2.9.18. Preparations for inhalation: aerodynamic assessment of fine particles. Strasbourg, France: Council of Europe; 2011. pp. 274–284.
4. *The United States Pharmacopeia*, (USP). Aerosols, Nasal Sprays, Metered Dose Inhalers, and Dry Powder Inhalers <601>. Rockville, MD, USA: The United States Pharmacopeial Convention; 2010.
5. Guo C, Gillespie S, Kauffman J, Doub W. Comparison of delivery characteristics from a combination metered-dose inhaler using the Andersen cascade impactor and the next generation pharmaceutical impactor. Journal of Pharmaceutical Sciences 2007;97(8):3321–3334.
6. Roberts DL, Romay FJ. Relationship of stage mensuration data to the performance of new and used cascade impactors. Journal of Aerosol Medicine and Pulmonary Drug Delivery 2005;18(4):396–413.
7. Zhou Y, Brasel TL, Kracko D, Cheng YS, Ahuja A, Norenberg JP, et al. Influence of impactor operating flow rate on particle size distribution of four jet nebulizers. Pharmaceutical Development and Technology 2007;12(4):353–359.
8. Stein SW. Aiming for a moving target: challenges with impactor measurements of MDI aerosols. International Journal of Pharmaceutics 2008;355(1–2):53–61.
9. Kamiya A, Sakagami M, Hindle M, Byron PR. Aerodynamic sizing of metered dose inhalers: an evaluation of the Andersen and next generation pharmaceutical impactors and their USP methods. Journal of Pharmaceutical Sciences 2004;93 (7):1828–1837.
10. Christopher D, Curry P, Doub B, Furnkranz K, Lavery M, Lin K, et al. Considerations for the development and practice of cascade impaction testing, including a

mass balance failure investigation tree. Journal of Aerosol Medicine and Pulmonary Drug Delivery 2003;16(3):235–247.

11. Berg E, Lamb P, Ali A, Dennis J, Tservistas M, Mitchell J. Assessment of the need to coat particle collection cups of the NGI to mitigate droplet bounce when evaluating nebuliser-produced droplets. Pharmeuropa Scientific Notes 2008;2008(1):2–5.

12. Rissler J, Asking L, Dreyer J. A methodology to study impactor particle reentrainment and a proposed stage coating for the NGI. Journal of Aerosol Medicine and Pulmonary Drug Delivery 2009;22(4):309–316.

13. Kamiya A, Sakagami M, Byron P. Cascade impactor practice for a high dose dry powder inhaler at 90 L/min: NGI versus modified 6-stage and 8-stage ACI. Journal of Pharmaceutical Sciences 2009;98(3):1028–1039.

14. Mitchell JP, Nagel MW, Wiersema KJ, Doyle CC. Aerodynamic particle size analysis of aerosols from pressurized metered-dose inhalers: comparison of Andersen 8-stage cascade impactor, next generation pharmaceutical impactor, and model 3321 aerodynamic particle sizer aerosol spectrometer. AAPS PharmSciTech 2003;4 (4):E54.

15. Marple VA, Roberts DL, Romay FJ, Miller NC, Truman KG, Van Oort M, et al. Next generation pharmaceutical impactor (a new impactor for pharmaceutical inhaler testing). Part I: design. Journal of Aerosol Medicine and Pulmonary Drug Delivery 2003;16(3):283–299.

16. Marple VA, Olson BA, Santhanakrishnan K, Mitchell JP, Murray SC, Hudson-Curtis BL. Next generation pharmaceutical impactor (a new impactor for pharmaceutical inhaler testing). Part II: archival calibration. Journal of Aerosol Medicine and Pulmonary Drug Delivery 2003;16(3):301–324.

17. Marple VA, Olson BA, Santhanakrishnan K, Roberts DL, Mitchell JP, Hudson-Curtis BL. Next generation pharmaceutical impactor: a new impactor for pharmaceutical inhaler testing. Part III: extension of archival calibration to 15 L/min. Journal of Aerosol Medicine and Pulmonary Drug Delivery 2005;17(4):335–343.

18. Thiel CG. Cascade impactor data and the lognormal distribution: nonlinear regression for a better fit. Journal of Aerosol Medicine and Pulmonary Drug Delivery 2003;15(4):369–378.

19. Svensson M, Pettersson G, Asking L. Mensuration and cleaning of the jets in Andersen cascade impactors. Pharmaceutical Research 2005;22(1):161–165.

20. Mitchell JP, Nagel MW. Cascade impactors for the size characterization of aerosols from medical inhalers: their uses and limitations. Journal of Aerosol Medicine and Pulmonary Drug Delivery 2004;16(4):341–377.

21. Keskinen J, Pietarinen K, Lehtimäki M. Electrical low pressure impactor. Journal of Aerosol Science 1992;23(4):353–360.

22. Marjamäki M, Keskinen J, Chen D-R, Pui DYH. Performance evaluation of the electrical low-pressure impactor (ELPI). Journal of Aerosol Science 2000;31 (2):249–261.

23. Anonymous. ELPI User Manual Version 3.13. Tampere, Finland: Dekati Ltd; 2001.

24. Baltensperger U, Weingartner E, Burtscher H, Keskinen J. Dynamic mass and surface area measurements. In Baron PA, Willeke K, editors. Aerosol Measurement: Principles, Techniques, and Applications. New York, NY: John Wiley & Sons, Ltd; 2001. pp. 387–418.

25. Van Guljik C, Schouten JM, Marijnissen JCM, Makkee M, Moulijn JA. Restriction for the ELPI in diesel particulate measurements. Journal of Aerosol Science 2001;32:1117–1130.

26. Van Guljik C, Marijnissen JCM, Makkee M, Moulijn JA. Oil-soaked sintered impactors for the ELPI in diesel particulate measurements. Journal of Aerosol Science 2003;34:635–640.

27. Tsukamoto Y, Goto Y, Odaka M. Continuous measurement of diesel particulate emissions by an electrical low-pressure impactor. SAE Technical Paper Series 2000-01-1138 2000: 143–148.

28. Ahlvik P, Ntziachristos L, Keskinen J, Virtanen A. Real time measurements of diesel particle size distribution with an electrical low pressure impactor. Society of Automotive Engineers Special Publication, SP-1335 (General Emissions) 1998: 215–234.

29. Shi JP, Mark D, Harrison RM. Characterization of particles from a current technology heavy-duty diesel engine. Environmental Science & Technology 2000;34:748–755.

30. Yokoi T, Sinzawa M, Matsumoto Y. Measurement repeatability improvement for particle number size distributions from diesel engines. JSAE Review 2001;22 (4):545–551.

31. Maricq MM, Podsiadlik DH, Chase RE. Gasoline vehicle particle size distributions: comparison of steady state, FTP, and US06 measurements. Environmental Science & Technology 1999;33(12):2007–2015.

32. Jetter JJ, Guo Z, McBrian JA, Flynn MR. Characterization of emissions from burning incense. The Science of the Total Environment 2002;295:51–67.

33. Yamada Y, Tokonami S, Yamasaki K. Applicability of the electrical low pressure impactor to size determination of aerosols attached to radon decay products. Review of Scientific Instruments 2005;76:065102.

34. Shi JP, Khan AA, Harrison RM. Measurements of ultrafine particle concentration and size distribution in the urban atmosphere. The Science of the Total Environment 1999;235:51–64.

35. Temesi D, Molnár A, Mészáros E, Feczkó T, Gelencsér A, Kiss G, et al. Size resolved chemical mass balance of aerosol particles over rural Hungary. Atmospheric Environment 2001;35:4347–4355.

36. Molnár A, Mészáros E. On the relation between the size and chemical composition of aerosol particles and their optical properties. Atmospheric Environment 2001;35:5053–5058.

37. Crampton M, Kinnersley R, Ayres J. Sub-micrometer particle production by pressurized metered dose inhalers. Journal of Aerosol Medicine 2004;17 (1):33–42.

38. Mikkanen P, Moisio M, Ristamäki J, Rönkkö T, Keskinen J, Korpiharju T. Measuring DPI charge properties using ELPI™. In Dalby RN, Byron PR, Peart J, Suman JD, Farr SJ, editors. Respiratory Drug Delivery IX. River Grove, IL: Davis Healthcare International; 2004. pp. 465–468.

39. Kwok PCL, Chan H-K. Electrostatic charge in pharmaceutical systems. In Swarbrick J, editor. Encyclopedia of Pharmaceutical Technology. New York, NY: Marcel Dekker; 2006. pp. 1535–1547.

40. Palonen M, Murtomaa M. Measurement of electrostatic charge of a pharmaceutical aerosol by ELPI, IFC, and FS. Poster presented at the International Society for Aerosols in Medicine 14th International Congress, Baltimore, MD. June 14–18, 2003.

41. Glover W, Chan H-K. Electrostatic charge characterization of pharmaceutical aerosols using electrical low-pressure impaction (ELPI). Journal of Aerosol Science 2004;35(6):755–764.

42. Kwok PCL, Chan H-K. Measurement of electrostatic charge of nebulised aqueous droplets with the electrical low pressure impactor. In Dalby RN, Byron PR, Peart J, Suman JD, Farr SJ, editors. Respiratory Drug Delivery IX. River Grove, IL: Davis Healthcare International; 2004. pp. 833–836.

43. Kwok PCL, Glover W, Chan H-K. Electrostatic charge characteristics of aerosols produced from metered dose inhalers. Journal of Pharmaceutical Sciences 2005;94 (12):2789–2799.

44. Kwok PCL, Collins R, Chan H-K. Effect of spacers on the electrostatic charge properties of metered dose inhaler aerosols. Journal of Aerosol Science 2006;37 (12):1671–1682.

45. Kotian R, Peart J. Influence of formulation components on inherent electrostatic properties of HFA propelled solution pMDIs. In Dalby RN, Byron PR, Peart J, Suman JD, Farr SJ, editors. Respiratory Drug Delivery X. River Grove, IL: Davis Healthcare International; 2006. pp. 947–950.

46. Kwok PCL, Noakes T, Chan H-K. Effect of moisture on the electrostatic charge properties of metered dose inhaler aerosols. Journal of Aerosol Science 2008;39 (3):211–226.

47. Telko MJ, Kujanpää J, Hickey AJ. Investigation of triboelectric charging in dry powder inhalers using electrical low pressure impactor (ELPI™). International Journal of Pharmaceutics 2007;336(2):352–360.

48. Kwok PCL, Chan H-K. Effect of relative humidity on the electrostatic charge properties of dry powder inhaler aerosols. Pharmaceutical Research 2008;25 (2):277–288.

49. Young PM, Sung A, Traini D, Kwok P, Chiou H, Chan H-K. Influence of humidity on the electrostatic charge and aerosol performance of dry powder inhaler carrier based systems. Pharmaceutical Research 2007;24(5):963–970.

7

Proteins, peptides, and controlled-release formulations for inhalation

Philip Chi Lip Kwok[1], Rania Osama Salama[2,3], and Hak-Kim Chan[2]

[1]*Department of Pharmacology and Pharmacy, LKS Faculty of Medicine, The University of Hong Kong, Hong Kong, China*
[2]*Advanced Drug Delivery Group, Faculty of Pharmacy, The University of Sydney, Sydney, Australia*
[3]*Faculty of Pharmacy, Alexandria University, Egypt*

7.1   Proteins and peptides for inhalation

Proteins are polymers consisting of amino acids covalently linked by peptide bonds. Peptides are small proteins composed of up to a few dozen amino acids [1]. The molecular weights of larger proteins may range from thousands to several millions of atomic masses, depending on the number of amino acids in the chain [2]. Due to their large sizes, absorption through the epithelial barriers in the gastrointestinal tract is slow [3]. Furthermore, proteins are rapidly degraded by digestive enzymes. Thus oral bioavailability is generally poor. The most common route of administration for pharmaceutical proteins is parenteral, including intravenous, subcutaneous, and intramuscular injections. However, the pulmonary route has also been used for protein delivery because the lungs possess a large surface area and an extensive vascular network for absorption [4]. From a practical viewpoint, inhalation delivery is noninvasive and more convenient than injections. This is especially of value for pharmaceutical proteins that require long-term administration, such as insulin for the systemic treatment of diabetes mellitus or recombinant human deoxyribonuclease (rhDNase), also known as dornase alfa, for the local treatment of cystic fibrosis.

Inhalation Drug Delivery: Techniques and Products, First Edition. Paolo Colombo, Daniela Traini, and Francesca Buttini.
© 2013 John Wiley & Sons, Ltd. Published 2013 by John Wiley & Sons, Ltd.

Many pharmaceutical proteins and peptides have been formulated as inhalation aerosols for local and systemic diseases (Table 7.1). The formulations are at various stages of development, ranging from Phase I studies to approved products. Exubera, for example, was a marketed insulin dry powder inhaler (DPI), but was withdrawn in October 2007 due to poor sales [5].

Table 7.1 Examples of proteins and peptides that have been applied as aerosols. Adapted from Cryan S-A. Carrier-based strategies for targeting protein and peptide drugs to the lungs. The AAPS Journal 2005;7(1):E20–E41. 7. Garcia-Contreras L, Smyth HDC. Liquid-spray or dry-powder systems for inhaled delivery of peptide and proteins? American Journal of Drug Delivery 2005;3(1):29–45

Local diseases	Compounds
α-1-antitrypsin deficiency	α-1-proteinase inhibitor
Asthma	Anti-IgE Mab
	Interleukin-1R
	Interleukin-4
	Lactoferrin
Antituberculosis vaccine	Muramyl dipeptide
Cancer/pneumocystis carnii	Interferon-γ
	Interleukin-2
Chronic bronchitis	Uridine triphosphate derivatives
Cystic fibrosis	rhDNase (approved)
	Targeted genetics adeno-associated virus for cystic fibrosis
Emphysema/cystic fibrosis	Alpha-1-antitrypsin
	Secretory leukoprotease inhibitor
Lung transplant	Cyclosporine A
Oxidative stress	Catalase
	Superoxide dismutase
Respiratory distress syndrome	Surfactant proteins (approved)

Systemic diseases	Compounds
Anemia	Erythropoietin
Anticoagulation	Heparin
Cancer	Interleukins
	Luteinizing hormone-releasing hormone
Diabetes insipidus	1-deaminocysteine-8-D-arginine vasopressin
Diabetes mellitus	Insulin (approved, but was later discontinued)
Endometriosis	Leuprolide
Growth hormone deficiency	Human growth hormone
Hemophilia	Factor IX
Multiple sclerosis	Interferon-β
Neutropenia	Recombinant human granulocyte-colony stimulating factor (rhG-CSF)
Osteoporosis	Calcitonin
	Parathyroid hormone
Viral infections	Interferon-α
	Ribavirin

7.1.1 Stability of proteins

The biological activity of proteins is strongly dependent on their molecular structure, which involves several organizational levels [1]. The primary structure is the amino acid sequence, which ultimately dictates the noncovalent interactions the molecule undergoes in forming higher-order structures. The secondary structure is the periodic spatial arrangement of the polypeptide chain backbone due to hydrogen bonding between the C=O and N—H groups. Alpha-helices and beta-sheets are typical secondary structures within a protein. The tertiary structure is the three-dimensional conformation of the whole molecule, including the positions of all amino acid side chains. Some proteins may consist of multiple peptide chains grouped together by noncovalent intermolecular interactions. The arrangement of the subunits relative to each other constitutes the quaternary structure. Alterations in the protein structure at any level may lead to a change or loss in biological activity.

Protein degradation can be chemical or physical in nature. Chemical degradation involves changes in the covalent bonds in the polypeptide chain, mainly through hydrolysis, isomerization, deamidation, oxidation, and disulfide bridging [8]. Hydrolysis is the cleavage of peptide bonds, usually under extreme acidic or alkaline conditions. However, even under neutrality the asparagine–proline and asparagine–glycine bonds are relatively labile [3]. Except for glycine, all naturally occurring amino acids are chiral and belong to the L-form. They can isomerize to the D-form under certain circumstances [8]. Deamidation is the transformation of the amide groups in asparagine and glutamine side chains into carboxylic acids, forming aspartic and glutamic acids, respectively [8]. The side chains of cysteine, methionine, tyrosine, tryptophan, and histidine are prone to oxidation [8]. The sulfur-containing side chains of cysteine and methionine can also form intra- or intermolecular covalent disulfide bridges [8]. All chemical degradations will disrupt the primary protein structure.

Physical degradation refers to changes in the noncovalent interactions within or between protein molecules. It can occur independently from chemical degradation and results in alterations in the higher-order structures (secondary and above). Common types of physical degradation include denaturation, aggregation, precipitation, and adsorption [8]. Denaturation is the unfolding of a molecule from its native conformation. It may or may not be reversible. Denatured proteins may also associate with each other to form molecular aggregates [8]. Precipitation is essentially aggregation on a macroscopic scale, where dissolved proteins come out from solution due to a reduction in solubility [8]. Protein molecules are surface-active and may adsorb on to various interfaces [8].

Owing to their complex structure, proteins are much more fragile than small drug molecules. The risks of chemical and physical degradation depend on many physico-chemical factors, such as temperature, pH, storage humidity, formulation constituents, delivery device, and manufacturing process, amongst others. These must be well studied and controlled to maintain the stability and efficacy of protein products.

7.1.2 Nebulizers and the AERx pulmonary delivery system

Protein nebule formulations are mainly aqueous solutions and are relatively easy to develop. The factors that may affect their stability are similar to those for general protein liquid formulations, such as ionic strength, pH, and the type and concentration of other dissolved compounds [9]. However, there are some potential problems that are unique to nebules. Atomizing air pressure, shear, recirculation, and progressive reduction of solution volume in jet nebulizers may lead to protein degradation [10]. Formulations without suitable protein stabilizers may also cause molecular adsorption to plastic surfaces [10]. The degradation rate of nebulized lactate dehydrogenase was found to increase with air pressure, nebulization time, and a low starting solution volume [11]. The aggregation and degradation rates of rhG-CSF also increased with nebulization time. However, these adverse effects were significantly reduced by the addition of polyethylene glycol (PEG) 1000 in the formulation [11]. This protective mechanism was attributed to the weak surface activity of PEG. The polymer may compete with the protein molecules to occupy the new liquid surfaces generated during nebulization, and consequently protect the proteins from degradation [11]. In another study, the stability of nebulized aviscumine, a dimeric protein, was improved by adding various surfactants, cryoprotectants, and buffer salts to the reconstitution medium [12–14].

High-energy vibrations and heat generated in ultrasonic nebulizers may denature proteins, as was observed on rhDNase [15,16]. Therefore, only jet nebulizers are approved for the administration of the marketed dornase alfa product Pulmozyme [17]. The formulation of Pulmozyme is very simple, with only sodium chloride, calcium chloride dihydrate, and water for injection as excipients [17,18]. This illustrates that not all protein nebule formulations require complex additives for stabilization. The stability of the protein of interest should be assessed on a case-by-case basis and excipients should be employed where appropriate.

The AERx pulmonary delivery system devised by Aradigm (California, USA) generates aerosols by mechanically extruding the liquid from a unit-dose reservoir through micron-sized orifices [19]. This device has been used for the delivery of a number of proteins, including rhDNase [20], insulin [21,22], and interleukin-4 receptor [23].

7.1.3 Metered-dose inhalers (MDIs)

The major issue in metered-dose inhaler (MDI) formulations of proteins is the conformational stability of the macromolecules in liquefied propellants, which are less polar than water. Fourier-transform Raman spectroscopy has been applied to investigate the secondary structure of hen egg lysozyme suspended in HFA 134a and HFA 227 [24]. Conformational data on the peptide backbone, C-C stretching, and disulfide bonds were obtained. The technique is simple and nondestructive, as the formulations can be analyzed in crimped glass vials [24].

Various proteins have been formulated as MDI suspensions [25,26]. Surfactants that are soluble in propellant were used to improve the suspendibility of the particles [25,26]. These included Triton X-100, Triton X-405, Laureth 9, Brij 30, Brij 97, Brij 98, Tween 80, Nonidet-40, and diethylene glycol monoethyl ether for HFA 134a [25,26]. An aqueous solution of the protein and surfactant was first freeze-dried to produce the required particles, which were subsequently suspended in HFA 134a alone or in a mixture of this propellant and dimethylether [25] or ethanol [26]. The propellant mixtures were found to improve the aerosol performance over the pure propellant. Kos Pharmaceuticals (Florida, USA) has developed an insulin MDI suspension in HFA 134a [27,28]. This formulation has been shown to provide comparable blood-glucose control to subcutaneous insulin injections in Type 2 diabetic patients [27].

Compared to MDI suspensions, it is more difficult to develop solutions for proteins due to the nonpolar nature of the propellants. Nevertheless, a stable HFA 134a solution containing leuprolide acetate was successfully formulated incorporating water and an undisclosed cosolvent [29].

7.1.4 Dry powder inhalers (DPIs)

Formulating protein powders for pulmonary delivery is challenging as it requires not only flowability and dispersibility of the powders but also stability of the proteins. To satisfy the latter requirement, proteins are usually formulated as amorphous glasses, which are, however, physically unstable and tend to crystallize with interparticulate bond formation and loss of powder dispersibility. In addition, the stability requirements limit the manufacturing processes that can be used for protein powder production.

During powder production, removal of water from the proteins can cause significant molecular conformational damage, which can lead to further chemical degradation including aggregation, deamidation, and oxidation during storage. Amorphous glassy excipients, mainly carbohydrates, have been widely used to stabilize proteins for inhalation, such as lactose for rhDNase [30,31]; trehalose, lactose, and mannitol for recombinant humanized anti-IgE monoclonal antibody (rhuMAbE25) [32]; and mannitol and raffinose for insulin [33]. Other suitable excipients include polymers (e.g. polyvinylpyrrolidone), proteins (e.g. human serum albumin), peptides (e.g. aspartame), amino acids (e.g. glycine), and organic salts (e.g. citrates). Although lactose has been widely used for inhalation products for small-molecule drugs, it may not be compatible with proteins. Being a reducing sugar, lactose is reactive toward lysine, and protein glycation has indeed been found in both rhDNase and rhuMAbE25 [32,34]. The exact mechanism for protein stabilization by excipients is as yet unclear. However, contributing factors may include (1) formation of a glassy state of the protein–excipient system, (2) hydrogen bonding between the protein and excipient molecules, (3) crystallinity of the excipients, and (4) residual water content. The diffusion rate and mobility of the protein molecules in the glassy state are much lower than those in the rubbery state. Therefore, any physicochemical reactions leading to chemical instability will be reduced [35]. Crystalline excipients such as mannitol are known to decrease the stability of proteins [36]. However, mannitol can be used in the amorphous form, for instance, in the presence of glycine [37]. Fourier-transform infrared spectroscopy offers information on protein secondary structures and has provided evidence for protein stabilization by hydrogen bonding [38]. Water promotes instability of proteins by enhancing molecular mobility [39], as shown by nuclear magnetic resonance (NMR) spectroscopy [40]. The crystalline or amorphous state of the excipients is crucial because it controls the distribution of water between the protein and the excipient in a powder [41].

While glassy materials are desirable for protein chemical stability, an immediate drawback is physical instability. Moisture uptake by fine particles of hydrophilic amorphous materials can be very fast, due to their large specific surface area and high energy state. For example, water uptake by spray-dried rhDNase with lactose induced crystallization that adversely affected powder dispersibility [30,31]. During crystallization, water acts as a plasticizer to lower the glass-transition temperature (T_g) (about 10 °C reduction per 1% water in sugar-containing formulations), which will enhance the molecular mobility required for nucleation if the T_g approaches the storage temperature [42]. It is thus important to keep the powder dry in order to maintain the high T_g, or to use excipients with a high T_g, or to store the powders at a low temperature. It has been proposed to store fragile glasses 50 °C below the T_g to minimize crystallization [43]. This method would be practical for room-temperature storage for amorphous materials with a T_g above 70 °C. It should be noted that the effect of moisture on powder dispersion can be instantaneous [44]. The hygroscopic effect can be reduced by using hydrophobic excipients, such as L-isoleucine [45].

Amorphous materials are usually cohesive, because they have higher surface energy and are more hygroscopic than the crystalline forms. Thus powder dispersibility is strongly dependent on the solid state. Most protein drugs are formulated with excipients for improved chemical stability, as discussed above. However, the distribution of protein and excipient(s) in a particle may not be uniform. When a protein–excipient solution droplet undergoes drying to form a particle, the outer surface tends to be enriched with proteins or macromolecules that are surface-active, while the small-molecule excipients diffuse rapidly into the particle core. Under special circumstances, small excipient molecules can crystallize on the particle surface [46]. Various formulation approaches have been employed to improve protein powder dispersibility. The particle size distribution can be controlled to decrease cohesion [46]. Proteins can be co-spray-dried with a suitable excipient to modify the surface energy and morphology; for example, rhDNase with sodium chloride [46]. Blending of protein drugs with inert carrier particles (e.g. lactose) can improve dispersibility [46]. Large (mean diameter $>5\,\mu m$) and small ($3-5\,\mu m$; PulmoSphere) porous particles containing proteins have excellent aerosol performance due to their low cohesion and small aerodynamic diameters [47–49]. Surface corrugation can also decrease cohesion by reducing the interparticulate contact areas. Wrinkled, nonporous, solid bovine serum albumin particles were reported to disperse significantly better than their nonwrinkled, spherical counterparts [50–52].

Inhalable protein powders can be produced by various methods, including spray-drying [49,53–55], spray freeze-drying [48], lyophilization followed by milling [56], and solvent precipitation with supercritical fluids [57–60]. Precipitation with regular antisolvents can be carried out using high-gravity controlled precipitation [61–63] or confined impinging jet-mixing [64,65]. Details on these processes are expounded in Chapter 4. Some issues regarding these techniques that are particularly relevant to protein formulations need to be noted. Spray-drying exposes the protein to mechanical shear and air–liquid interfacial denaturation. The hot air for drying also subjects the protein to thermal stress and denaturation. Aggregation of spray-dried recombinant human growth hormone was suppressed by adding Zn^{2+} ions and surfactant polysorbate 20 [53,54]. Compared to spray-drying, spray freeze-drying produces porous and light particles with superior dispersibility, and the production yield is almost 100%. However, this process is more time-consuming and expensive. Gas-jet-milling of lyophilized proteins can cause contamination and inactivation. Thus abrasive-resistant mills using high-purity nitrogen and milling stabilizers such as human serum albumin and sorbitol are required [56]. Supercritical carbon dioxide is a good antisolvent for precipitation because it is nontoxic, economical, and has a low critical temperature of 31.1 °C for operation. However, due to its nonpolar nature, it is not readily miscible with water. Special nozzles have been employed to enhance mixing with aqueous protein solutions [59,60]. Alternatively, supercritical carbon dioxide modified with ethanol can be used as an antisolvent [57,58]. Since carbon dioxide is acidic, the pH of the mixture should be controlled to avoid protein degradation. Excipients such as PEG and self-assembling small organic molecules can be combined with proteins to form microparticles by precipitation [66–68].

7.2 Controlled-release formulations for inhalation

Two significant drawbacks of current inhalation therapies are the short duration of action and the need to deliver drugs at least three to four times daily [69]. Despite the increasing number of inhalation formulations for the treatment of respiratory illnesses, no controlled-release inhalation system exists to date. Maximizing the concentration/dose ratio in the lung tissue by delivering the medicament to the upper airways will potentially decrease systemic exposure [70]. In addition, by promoting systemic absorption by delivering the inhalation dose to the lower respiratory tract, the delivery of systemic agents becomes feasible [71].

Many drug molecules have been investigated for their potential as local controlled-release agents in the lung, including antibacterials [72,73], antivirals [74,75], antifungals [76–78], cytotoxic agents [73], and immunosuppressives [79]. In addition, by prolonging the presence of β_2-agonists (as asthma is influenced by circadian rhythm), the exacerbation of asthma during sleep may potentially be reduced [80]. Insulin, rhDNase, α-1-antitrypsin, DNA vaccines, interferons, calcitonin, human growth hormone, parathyroid hormone, vaccines, gene therapy, and leuprolide are just a few examples of systemic molecules that are candidates for systemic controlled-release inhalation therapies [81–84]. Another example is morphine, a small painkiller molecule, which would be extremely advantageous for cancer patients or for postoperative pain management if formulated in a controlled-release inhalation formulation [85].

7.2.1 Challenges facing controlled-release inhalation therapy

The controlled-release formulations have complex challenges which add to the requirements of formulating any inhalation therapy. Not only should the particles be inhalable (aerodynamic diameter $<5\,\mu m$), but a release-modifying excipient is needed for controlled release of the drug after delivery. Both of these steps are challenging: the production of micron-sized particles is difficult, since with a reduction in size there is a significant increase in surface area-to-mass ratio [86–88]; subsequently, with an increase in surface area, it becomes more difficult to produce a controlled-release profile and incorporate effective release agents.

The physiology of the lung and its impact on resident particles is another hurdle. The deposited particles will be primarily removed by the mucociliary escalator towards the pharynx and ultimately deposited in the gastrointestinal tract within 24 hours [89], if the aerodynamic particle size of a formulation is in the range $2.5–6.0\,\mu m$ (as for local therapy). Moreover, some of these particles may be absorbed though the epithelium in this region into the blood or the lymphatic system [90].

Alternatively, insoluble particles deposited in the alveolar sac are subjected to clearance mainly via alveolar macrophages. These exist in numbers equivalent to five to seven macrophages per alveolus [91]. They engulf and enzymatically degrade foreign particulate matter or microorganisms, and migrate the results to the mucociliary escalator or lymph tissue [92] in a time frame of weeks to months [89]. This is the case when particles have an aerodynamic diameter $<2.5\,\mu m$, as for systemic delivery, where the particles will be primarily deposited in the alveoli [93]. In Section 7.2.2, the different strategies used to produce controlled-release formulations for inhalation are highlighted.

7.2.2 Production of inhalable controlled-release formulations

Biodegradable excipient-based matrices

These matrices range from synthetic to natural materials. Examples include biocompatible synthetic polymers such as polylactic co-glycolic acid (PLGA), polylactic acid (PLA), PEG, and polyvinyl alcohol (PVA), and natural polymers or proteins such as chitosan and albumin.

These systems depend on the poor solubility of the polymer or drug–polymer interaction to affect the drug release. When drug molecules are incorporated into these biodegradable matrices and deposited at the air–liquid interface in the lung, they slowly release the active drug. The drug then passively diffuses out of the particulate to be slowly released into the systemic circulation [75,94,95].

PLGA and PLA have been the most commonly reported polymers utilized for potential respiratory sustained-release systems. Although they are usually prepared via double emulsion formation followed by solvent evaporation [96,97], spray-drying fluidized bed granulation was used by Yamamoto et al. for the production of PLGA nanocomposite granules [98]. In this study, insulin-loaded nanospheres with mannitol showed significantly decreased blood glucose levels as well as prolonged pharmacological effects due to the preferable inhalation performance and gradual release of insulin from nanospheres in the nanocomposite.

Production of porous, low-density, respirable particles were also achieved by spray-drying active ingredients dissolved in ethanolic solution with other excipients [99–101]. PLGA and PLA were used to form nonbrittle, shell-like, large particles utilizing this method for controlled-release powder production. Edwards et al. produced large-size ($>5\,\mu$m), low-density ($<0.4\,$g/cm^3) particles of both insulin and testosterone with PLGA and PLA [47]. The porous particles were inhaled deep into the lungs and were detected for longer when compared with the conventional particles (96 hours compared to 4 hours, respectively).

Large porous particles were also prepared using a supercritical CO_2 treatment process. This method was used to prepare large porous deslorelin-PLGA particles with reduced residual solvent content, retained deslorelin integrity, sustained release, and reduced cellular uptake [102].

Highly porous large PLGA microparticles were recently produced using ammonium bicarbonate as an effervescent porogenic agent. These resultant particles had suitable aerodynamic properties and good encapsulation efficiency, in addition to a reduced macrophage uptake and a sustained release pattern of doxorubicin [103].

PVA has also been studied as a potential biocompatible polymer for pulmonary delivery. PVA was found to improve the aerosolization efficiency of spray-dried disodium cromoglycate (DSCG) and to prolong the release of the active drug from the microparticles [104,105]. Bovine serum albumin microparticles for pulmonary inhalation were also produced using PVA, and similarly showed prolonged release profiles [106]. Interestingly, another study by Liao et al. showed that the use of PVA in the formulation of pressurized MDIs physically stabilized suspensions of test proteins by limiting the potential for irreversible aggregation [107].

It is important to note that in spite of the popularity and the success of using polymeric release-modifying agents, the long residence time due to slow degradation might lead to pulmonary accumulation of these polymers, especially with daily administration [108,109]. Furthermore, pulmonary administration of PLA microspheres to rabbits was associated with raised neutrophil count, inflammation at sites close to microparticle deposition, and hemorrhage [110]. In addition, in cell-based toxicity screening, PLGA and PLA significantly reduced cell viability when compared to lipid-based particles [111]. A recent comparative study of a series of potential polymers has shown that PLGA had the greatest toxicity of the polymers when studied on the Calu-3 monolayers [112], in comparison with a preliminary A549 cell toxicity study, which indicated that PVA had limited effect on cell viability after 24 hours' exposure [106]. While the same study by Sivadas et al. showed that hydroxypropyl cellulose had high delivery efficiency, sodium hyaluronate and chitosan showed low toxicity and controlled-release behavior, and ovalbumin and chitosan had improved systemic delivery of a model protein-loaded particle system prepared by spray-drying.

Avoidance of mucociliary clearance and potentially prolonged residence time through physical adsorption may be achieved by the use of mucoadhesive polymers such as chitosan [113,114] and hydroxypropyl cellulose [115]. Chitosan is also an absorption enhancer as it has been reported to loosen the tight junctions between the epithelial cells and thus facilitate absorption [116,117]. Surface modification of PLGA nanospheres using chitosan as a mucoadhesive coat has been reported to lead to slower elimination from the lungs and to enhance the absorption of the active drug compared to unmodified PLGA nanospheres [117]. Co-spray-dried chitosan formulations containing either soluble terbutaline sulfate or insoluble beclomethasone dipropionate drug molecules had significantly longer release rates than the drug alone [118,119]. Formulations containing chitosan and beclamethasone dipropionate showed in vitro respiratory efficiencies >43% and up to 12 hours' release in 1000 mL phosphate buffer [118]. Furthermore, the release of encapsulated insulin could be controlled by incorporating a negatively charged phospholipid in a film coating chitosan nanoparticles [120].

To overcome potential toxicity and accumulation of these polymers in the lung, the use of natural protein carriers may be a suitable alternative. Albumin incorporation in the formulation of DPIs is expected to prolong the release of particles in the alveoli [121]. Albumin microspheres were reported to deliver tetrandine [122] and were found to provide greater drug concentrations in the lungs when compared to conventional oral dosing. Kwon et al. prepared bovine serum albumin porous microparticles using sucrose ethyl isobutylate as an additive for sustained protein release and a cyclodextrin derivative as a porogen; the in vitro release was observed up to 7 days [123]. In another study by Li et al., albumin microspheres entrapping an antibacterial drug showed in vitro sustained-release profiles for over 12 hours [121]. Addition of albumin may also have a dramatic effect on the morphology and fine-particle fraction of the spray-dried powders [124].

Examples of natural gums being studied as potential release-modifying agents are xanthan and locust bean gums. These gums were shown to reduce the release of salbutamol, indicating that a strong gel was formed by the intermolecular interaction between xanthan and locust bean gums [125].

Increasing the total molecular weight of the system and hence retarding the rate of absorption of the active ingredient across the epithelium of the alveoli can be achieved by forming conjugates using large-molecular-mass polymers or proteins [94]. PEG is a macromolecule that has been conjugated with proteins (e.g. asparaginase) [94,126]. These conjugates can be conveniently fabricated to present a variety of ligands at the distal ends of spacer PEG chains [127,128]. Using calcium phosphate–PEG particles, the bioavailability and duration of action of insulin were enhanced when administered to the lungs of rats compared to conventional subcutaneous delivery [129].

It is important to note that attaching polymeric materials to proteins requires careful understanding of the positions of the functional groups so as not to block the active protein site and provoke loss of activity.

Molecular dispersions

A liposomal vesicle's size—ranging from around 20 nm to several microns—surface charge, number of bilayers, and method of preparation are the major parameters controlling its half-life and the extent of drug entrapment [130]. Furthermore, since liposomes can be formulated from different types of lipid, it is possible to produce a wide range of physicochemical properties that can accommodate a wide range of encapsulated drug molecules. Liposomal sizes of 50–200 nm are considered to be optimal for avoiding phagocytosis by macrophages while still being useful for encapsulating drugs [131]. Liposomal and phospholipid-based liquid formulations have been used for the treatment of neonatal respiratory distress syndrome and seasonal asthma [132], thus making them relatively safe vehicles [133] for controlled-release applications, in comparison with the biodegradable polymeric matrices.

In addition, different surface charges and bilayer fluidities, depending on the production conditions and chemical composition, can be manipulated for a desired physiological effect. Cationic liposomes have been used to bind negatively charged DNA and fuse to cell membranes for the treatment of cystic fibrosis [134,135]. Such an approach, in spite of decreasing transfection efficiencies, avoids the immunogenic response and risks associated with the use of viral vectors [136].

Liposomal formulations have generally been found to reduce systemic side effects and improve clinical effects. The effect of the liposomal encapsulation of bronchodilators (sympathomimetic amines), anti-asthma drugs (sodium cromoglycate), antimicrobial agents (amikacin, benzylpenicillin, oxytocin), antiviral agents (enviroxine), cytotoxic agents (β-cytosine arabinoside), and antioxidant agents (catalase, superoxide dismutase, glutathione) has been summarized by Zeng et al. [75]. When insulin is encapsulated in liposomes, its absorption from the lungs is enhanced and prolonged [75].

Significantly higher drug levels and prolonged drug retention in the respiratory tract have been observed in a liposomal formulation by jet nebulization when compared to free ciprofloxacin [137]. In this study, aerosol inhalation of liposome-encapsulated ciprofloxacin, given either prophylactically or therapeutically, provided complete protection to mice against a pulmonary lethal infection model of *F. tularensis*. In contrast, ciprofloxacin given in its free form was ineffective.

More recently, Stark et al. encapsulated a vasoactive intestinal peptide into unilamellar liposomes [138]. They found that sterically stabilized liposomal formulations have the potential to enhance the life-span and biological activity of peptide drugs in the metabolic environment of the lung. Chono et al. indicated that, according to pharmacokinetic/pharmacodynamic analysis, the pulmonary administration of mannosylated ciprofloxacin-liposomes exhibited potent antibacterial effects against many bacteria tested and could be useful against intracellular parasitic infections [139].

A modification of the liposomal system is agglomerated vesicle technology. In vivo cleavage of the agglomerated liposomes, which are used as core nanoparticles, results in controlled release of the drug [127,128,140].

Solid lipid particles

Formulating liposomes into dry powders [141,142] may be a good alternative for overcoming the high production cost, disruption, and loss of entrapped medicaments during storage or nebulization which remain significant challenges in the production and stability of liposomes [143,144]. These systems are more chemically and physically stable than liposomes, have good entrapment yields, use inexpensive carriers, and avoid toxic solvent residues [145]. Among the limited number of reports available in this field, Jaspart et al. pointed to salbutamol acetonide-loaded solid lipid nanoparticles of glyceryl behenate [146]; these formulations showed delayed release profiles in vitro when compared to the free drug or the physical mixtures, with no significant inflammatory airway response observed in similar systems after intratracheal administration in rats [147].

Coating or encapsulating drug particles in a lipid outer shield was found to decrease microparticle uptake by macrophages. The uptake of PLGA microparticles by cultured macrophages was found to be significantly reduced when dipalmitoyl-phosphatidylcholine was incorporated into the formulation [148]. Furthermore, the in vitro release of the incorporated drugs was also retarded in this case. Cook et al. demonstrated reduced in vitro release rates of terbutaline sulfate when the primary particles were coated with a hydrogenated palm oil and dipalmitoylphosphatidyl-choline lipid system [141]. Similarly, chitosan nanoparticles coated with a lipid film of dipalmitoylphosphatidylcholine and dimyristoylphosphatidylcholine were successful in controlling the release of insulin from the nanoparticles under in vitro conditions [120].

Viscous systems

Forming a gel interface for the passive transport of the uniformly dispersed drug through the gel matrix upon deposition in the lung is the principle of this approach. Pulmonary absorption of 5(6)-carboxyfluorescein was regulated using 5% gelatine, 1% PVA, 1% hydroxypropyl cellulose, 1% methyl cellulose 400, or 1% hyaluronic acid [149]. Intratracheal administration of these aqueous solutions had a significant effect on drug plasma levels in rats. On the other hand, the excipients' effect on the release profile appeared to be drug-specific, since the release rate of fluorescein isothiocyanate-labeled dextrans was not regulated by gelatine or PVA. Similarly, iota- and kappa-carrageenans have been shown to be potential release-modifying polysaccharide gels. A modified absorption rate of theophylline and flutecasone propionate was observed when the polymers were utilized in solutions <5% w/v, with no evident damaging or inflammatory effect on lung tissue [150].

It is interesting to note that all these approaches have shown successful release profiles either in vitro (i.e. using different conventional and modified dissolution methodologies) or in vivo (i.e. in murine models). However, the methodologies and approaches for evaluating these systems are diverse. Salama et al. have compared different in vitro methods in use to evaluate the release profiles of controlled-release DPIs and found significant differences with different approaches [105]. Without a standardized method of assessing and testing controlled-release formulations for inhalation, it becomes quite difficult to accurately compare methods of production and their effectiveness. Although many challenges exist, controlled-release formulations to the lung have not yet reached their full potential and are still underappreciated.

References

1. Campbell MK. Biochemistry. Orlando, FL: Harcourt Brace & Company; 1999.

2. Bailey PSJr, Bailey CA. Organic Chemistry: A Brief Survey of Concepts and Applications. Upper Saddle River, NJ: Prentice-Hall International; 1995.

3. Crommelin D, van Winden E, Mekking A. Delivery of pharmaceutical proteins. In: Aulton ME, editor. Pharmaceutics: The Science of Dosage Form Design. Edinburgh: Churchill Livingstone; 2002. pp. 544–553.

4. Qiu Y, Gupta PK, Adjei AL. Absorption and bioavailability of inhaled peptides and proteins. In: Adjei AL, Gupta PK, editors. Inhalation Delivery of Therapeutic Peptides and Proteins. New York, NY: Marcel Dekker; 1997. pp. 89–131.

5. Bailey CJ, Barnett AH. Why is Exubera being withdrawn? British Medical Journal 2007;335:1156.

6. Cryan S-A. Carrier-based strategies for targeting protein and peptide drugs to the lungs. AAPS Journal 2005;7(1):E20–E41.

7. Garcia-Contreras L, Smyth HDC. Liquid-spray or dry-powder systems for inhaled delivery of peptide and proteins? American Journal of Drug Delivery 2005;3 (1):29–45.

8. Manning MC, Patel K, Borchardt RT. Stability of protein pharmaceuticals. Pharmaceutical Research 1989;6(11):903–918.

9. Arakawa T, Prestrelski SJ, Kenney WC, Carpenter JF. Factors affecting short-term and long-term stabilities of proteins. Advanced Drug Delivery Reviews 1993;10:1–28.

10. Gupta PK, Adjei AL. Therapeutic inhalation aerosols. In: Adjei AL, Gupta PK, editors. Inhalation Delivery of Therapeutic Peptides and Proteins. New York, NY: Marcel Dekker; 1997. pp. 185–234.

11. Niven RW, Ip AY, Mittelman SD, Farrar C, Arakawa T, Prestrelski SJ. Protein nebulization: I. Stability of lactate dehydrogenase and recombinant granulocyte-colony stimulating factor to air-jet nebulizaion. International Journal of Pharmaceutics 1994;109:17–26.

12. Steckel H, Eskander F, Witthohn K. Effect of excipients on the stability and aerosol performance of nebulised aviscumine. Journal of Aerosol Medicine 2003;16:417–432.

13. Steckel H, Eskander F, Witthohn K. Effect of cryoprotectants on the stability and aerosol performance of nebulized aviscumine, a 57-kDa protein. European Journal of Pharmaceutics and Biopharmaceutics 2003;56:11–21.

14. Steckel H, Eskander F, Witthohn K. The effect of formulation variables on the stability of nebulised aviscumine. International Journal of Pharmaceutics 2003;257:181–194.

15. Phipps PR, Gonda I. Droplets produced by medical nebulizers: some factors affecting their size and solute concentration. Chest 1990;97:1327–1332.

16. Cipolla DC, Clark AR, Chan H-K, Gonda I, Shire SJ. Assessment of aerosol delivery systems for recombinant human deoxyribonuclease. STP Pharma Sciences 1994;4:50–62.

17. Roche, Products. Pulmozyme® Consumer Medicine Information. Dee Why, NSW, Australia: Roche Products Pty Ltd; 2008.

18. Gonda I. Deoxyribonuclease inhalation. In: Adjei AL, Gupta PK, editors. Inhalation Delivery of Therapeutic Peptides and Proteins. New York, NY: Marcel Dekker; 1997. pp. 355–365.

19. Rubsamen R. Novel aerosol peptide drug delivery systems. In: Adjei AL, Gupta PK, editors. Inhalation Delivery of Therapeutic Peptides and Proteins. New York, NY: Marcel Dekker; 1997. pp. 703–731.

20. Mudumba S, Khossravi M, Yim D, Rossi T, Pearce D, Hughes M, et al. Delivery of rhDNase by the AERx® pulmonary delivery system. In: Dalby RN, Byron PR, Farr S, Peart J, editors. Respiratory Drug Delivery VII. Raleigh, NC: Serentec Press; 2000. pp. 329–332.

21. Henry RR, Mudaliar SRD, Howland WC III, Chu N, Kim D, An B, et al. Inhaled insulin using the AERx Insulin Diabetes Management System in healthy and asthmatic subjects. Diabetes Care 2003;26:764–769.

22. Farr S, Reynolds D, Nat A, Srinivasan S, Roach M, Jensen S, et al. Technical development of AERx® Diabetes Management System: essential characteristics for diabetes treatment with pulmonary insulin. In: Dalby RN, Byron PR, Peart J, Farr S, editors. Respiratory Drug Delivery VIII. Raleigh, NC: Davis Horwood International; 2002. pp. 51–60.

23. Sangwan S, Agosti JM, Bauer LA, Otulana BA, Morishige RJ, Cipolla DC, et al. Aerosolized protein delivery in asthma: gamma camera analysis of regional deposition and perfusion. Journal of Aerosol Medicine 2001;14:185–195.

24. Quinn ÉÁ, Forbes RT, Williams AC, Oliver MJ, McKenzie L, Purewal TS. Protein conformational stability in the hydrofluoroalkane propellants tetrafluoroethane and heptafluoropropane analysed by Fourier transform Raman spectroscopy. International Journal of Pharmaceutics 1999;186:31–41.

25. Brown AR, George DW. Tetrafluoroethane (HFC 134A) propellant-driven aerosols of proteins. Pharmaceutical Research 1997;14:1542–1547.

26. Williams RO III, Liu J. Formulation of a protein with propellant HFA 134a for aerosol delivery. European Journal of Pharmaceutical Sciences 1998;7:137–144.

27. Hausmann M, Dellweg S, Osborn C, Heinemann L, Buchwald A, Rosskamp R, et al. Inhaled insulin as adjunctive therapy in subjects with type 2 diabetes failing oral agents: a controlled proof-of-concept study. Diabetes, Obesity and Metabolism 2006;8:574–580.

28. Adjei AL, Genova P, Zhu Y, Sexton F. Method of treating a systemic disease. US patent 7056494. 2006.

29. Brambilla G, Berrill S, Davies RJ, Ganderton D, George SC, Lewis DA, et al. Formulation of leuprolide as an HFA solution pMDI. Journal of Aerosol Medicine 2003;16:209.

30. Chan H-K, Gonda I. Solid state characterization of spray-dried powders of recombinant human deoxyribonuclease (rhDNase). Journal of Pharmaceutical Sciences 1998;87(5):647–654.

31. Clark AR, Dasovich N, Gonda I, Chan H-K. The balance between biochemical and physical stability for inhalation protein powders: rhDNase as an example. In: Dalby RN, Byron PR, Farr S, editors. Resiratory Drug Delivery V. Buffalo Grove, IL: Interpharm Press; 1996. pp. 167–174.

32. Andya JD, Maa Y-F, Constantino HR, Nguyen P-A, Dasovich N, Sweeney TD, et al. The effect of formulation excipients on protein stability and aerosol performance of spray-dried powders of a recombinant humanized anti-IgE monoclonal antibody. Pharmaceutical Research 1999;16:350–358.

33. Patton JS, Foster L, Platz RM. Methods and compositions for pulmonary delivery of insulin. US patent 5997848. 1999.

34. Quan CP, Wu S, Dasovich N, Hsu C, Patapoff T, Canova-Davis E. Susceptibility of rhDNase I to glycation in the dry-powder state. Analytical Chemistry 1999;71:4445–4454.

35. Frank F. Long-term stabilization of biologicals. Biotechnology 1994;12:253–256.

36. Izutsu K-I, Yoshioka S, Terao T. Decreased protein-stabilizing effects of cryoprotectants due to crystallization. Pharmaceutical Research 1993;10:1232–1237.

37. Pikal MJ, Dellerman KM, Roy ML, Riggin RM. The effects of formulation variables on the stability of freeze-dried human growth hormone. Pharmaceutical Research 1991;8:427–436.

38. Carpenter JF, Prestrelski SJ, Dong A. Application of infrared spectroscopy to development of stable lyophilized protein formulations. European Journal of Pharmaceutics and Biopharmaceutics 1998;45:231–238.

39. Hageman MJ. Sorption and solid-state stability of proteins. In: Ahern TJ, Manning MC, editors. Stability of Protein Pharmaceuticals, Part A: Chemical and Physical Pathways of Protein Degradation. New York, NY: Plenum Press; 1992. pp. 273–309.

40. Separovic F, Lam YH, Ke X, Chan H-K. A solid-state NMR study of protein hydration and stability. Pharmaceutical Research 1998;15:1816–1821.

41. Chan H-K, Au-Yeung K-L, Gonda I. Development of a mathematical model for the water distribution in freeze-dried solids. Pharmaceutical Research 1999;16:660–665.

42. Roos Y, Karel M. Plasticizing effect of water on thermal behavior and crystallization of amorphous food models. Journal of Food Science 1991;56:38–43.

43. Hancock BC, Shamblin SL, Zografi G. Molecular mobility of amorphous pharmaceutical solids below their glass transition temperatures. Pharmaceutical Research 1995;12:799–806.

44. Chew NYK, Chan H-K. Effect of humidity on the dispersion of dry powers. In: Dalby RN, Byron PR, Farr S, Peart J, editors. Respiratory Drug Delivery VII. Raleigh, NC: Serentec Press; 2000. pp. 615–618.

45. Yamashita C, Nishibayashi T, Akashi S, Toguchi H, Odomi M. A novel formulation of dry powder for inhalation of peptides and proteins. In: Dalby RN, Byron PR, Farr S, editors. Respiratory Drug Delivery V. Buffalo Grove, IL: Interpharm Press; 1996. pp. 483–486.

46. Chan H-K, Clark A, Gonda I, Mumenthaler M, Hsu C. Spray dried powders and powder blends of recombinant human deoxyribonuclease (rhDNase) for aerosol delivery. Pharmaceutical Research 1997;14:431–437.

47. Edwards DA, Hanes J, Caponetti G, Hrkach J, Ben-Jebria A, Eskew ML, et al. Large porous particles for pulmonary drug delivery. Science 1997;276 (5320):1868–1872.

48. Maa Y-F, Nguyen P-A, Sweeney TD, Shire SJ, Hsu CC. Protein inhalation powders: spray drying vs spray freeze drying. Pharmaceutical Research 1999;16:249–254.

49. Bot AI, Tarara TE, Smith DJ, Bot SR, Woods CM, Weers JG. Novel lipid-based hollow-porous microparticles as a platform for immunoglobulin delivery to the respiratory tract. Pharmaceutical Research 2000;17:275–283.

50. Chew NYK, Chan HK. Use of solid corrugated particles to enhance powder aerosol performance. Pharmaceutical Research 2001;18(11):1570–1577.

51. Chew NYK, Tang P, Chan HK, Raper JA. How much particle surface corrugation is sufficient to improve aerosol performance of powders? Pharmaceutical Research 2005;22(1):148–152.

52. Adi S, Adi H, Tang P, Traini D, Chan H-K, Young PM. Micro-particle corrugation, adhesion and inhalation aerosol efficiency. European Journal of Pharmaceutical Sciences 2008;35:12–18.

53. Mumenthaler M, Hsu CC, Pearlman R. Feasibility study on spray-drying protein pharmaceuticals: recombinant human growth hormone and tissue-type plasminogen activator. Pharmaceutical Research 1994;11:12–20.

54. Maa Y-F, Nguyen P-AT, Hsu SW. Spray-drying of air-liquid interface sensitive recombinant human growth hormone. Journal of Pharmaceutical Sciences 1998;87:152–159.

55. Vanbever R, Mintzes JD, Wang J, Nice J, Chen D, Batycky R, et al. Formulation and physical characterization of large porous particles for inhalation. Pharmaceutical Research 1999;16:1735–1742.

56. Platz RM, Ip A, Whitham CL., Process for preparing micronized polypeptide drugs. US patent 5354562. 1994.

57. Bustami RT, Chan H-K, Dehghani F, Foster NR. Generation of micro-particles of proteins for aerosol delivery using high pressure modified carbon dioxide. Pharmaceutical Research 2000;17(11):1360–1366.

58. Bustami RT, Chan H-K, Foster NR. Aerosol delivery of protein powders processed by supercritical fluid technology. In: Dalby RN, Byron PR, Farr S, Peart J, editors. Respiratory Drug Delivery VII. Raleigh, NC: Serentec Press; 2000. pp. 611–614.

59. Sievers RW, Sellers SP, Kusek KD, Glark GS, Korte BJ. Fine-particle formation using supercritical carbon dioxide-assisted aerosolization and bubble drying. In: Proceedings of the 218th ACS National Meeting. New Orleans, LA; American Chemical Society: 1999.

60. Sloan R, Hollowood HE, Hupreys GO, Ashraf W, York P. Supercritical fluid processing: preparation of stable protein particles. In: Proceedings of the Fifth Meeting of Supercritical Fluids. Nice, France; 1998.

61. Chiou H, Li L, Hu T, Chan HK, Chen JF, Yun J. Production of salbutamol sulfate for inhalation by high-gravity controlled antisolvent precipitation. International Journal of Pharmaceutics 2007;331(1):93–98.

62. Chen J-F, Zhou M-Y, Shao L, Wang Y-Y, Yun J, Chew NYK, et al. Feasibility of preparing nanodrugs by high-gravity reactive precipitation. International Journal of Pharmaceutics 2004;269:267–274.

63. Hu T, Chiou H, Chan H-K, Chen J-F, Yun J. Preparation of inhalable salbutamol sulfate using reactive high gravity controlled precipitation. Journal of Pharmaceutical Sciences 2008;97:944–949.

64. Chiou H, Chan H-K, Prud'homme RK, Raper JA. Evaluation on the use of the confined liquid impinging jets for the synthesis of nanodrug particles. Drug Development and Industrial Pharmacy 2008;34:59–64.

65. Chiou H, Chan H-K, Heng D, Prud'homme RK, Raper JA. A novel production method for inhalable cyclosporine A powders by confined liquid impinging jet precipitation. Journal of Aerosol Science 2008;39:500–509.

66. Steiner SS, Pfützner A, Wilson BR, Harzer O, Heinemann L, Rave K. Technosphere/insulin: proof of concept study with a new insulin formulation for pulmonary delivery. Experimental and Clinical Endocrinology & Diabetes 2002;110: 17–21.

67. Lian H, Steiner SS, Sofia RD, Woodhead JH, Wolf HH, White HS, et al. A self-complementary, self-assembling microsphere system: application for intravenous delivery of the antiepileptic and neuroprotectant compound felbamate. Journal of Pharmaceutical Sciences 2000;89:867–875.

68. Brown LR, Rashba-Step J, Scott TL, Qin Y, Rulon PW, McGeehan J, et al. Pulmonary delivery of novel insulin microspheres. In: Dalby RN, Byron PR, Peart J, Farr S, editors. Respiratory Drug Delivery VIII. Raleigh, NC: Davis Horwood International; 2002. pp. 431–434.

69. Byron PR. Prediction of drug residence times in regions of the human respiratory tract following aerosol inhalation. Journal of Pharmaceutical Sciences 1986;75 (5):433–438.

70. Bennett WD, Brown JS, Zeman KL, Hu S-C, Scheuch G, Sommerer K. Targeting delivery of aerosols to different lung regions. Journal of Aerosol Medicine 2002;15(2):179–188.

71. Jain KK. Drug delivery systems—an overview. In: Jain KK, editor. Drug Delivery Systems. Totowa, NJ: Humana Press; 2008. pp. 1–50.

72. Poyner EA, Alpar HO, Almeida AJ, Gamble MD, Brown MRW. A comparative study on the pulmonary delivery of tobramycin encapsulated into liposomes and PLA microspheres following intravenous and endotracheal delivery. Journal of Controlled Release 1995;35(1):41–48.

73. Schreier H, Gonzalez-Rothi RJ, Stecenko AA. Pulmonary delivery of liposomes. Journal of Controlled Release 1993;24(1–3):209–223.

74. Taylor KM, Newton JM. Liposomes for controlled delivery of drugs to the lung. Thorax 1992;47(4):257–259.

75. Zeng XM, Martin GP, Marriott C. The controlled delivery of drugs to the lung. International Journal of Pharmaceutics 1995;124(2):149–164.

76. Gilbert BE. Liposomal aerosols in the management of pulmonary infections. Journal of Aerosol Medicine 1996;9(1):111–122.

77. Ruijgrok EJ, Vulto AG, Van Etten EW. Aerosol delivery of amphotericin B desoxycholate (Fungizone) and liposomal amphotericin B (AmBisome): aerosol characteristics and in-vivo amphotericin B deposition in rats. Journal of Pharmacy and Pharmacology 2000;52(6):619–627.

78. Alvarez CA, Wiederhold NP, McConville JT, Peters JI, Najvar LK, Graybill JR, et al. Aerosolized nanostructured itraconazole as prophylaxis against invasive pulmonary aspergillosis. Journal of Infection 2007;55(1):68–74.

79. Behr J, Zimmermann G, Baumgartner R, Leuchte H, Neurohr C, Brand P, et al. Lung deposition of a liposomal cyclosporine a inhalation solution in patients after lung transplantation. Journal of Aerosol Medicine and Pulmonary Drug Delivery 2009;22(2):1–9.

80. Smolensky MH, D'Alonzo GE, Kunkel G, Barnes PJ. Day-night patterns in bronchial patency and dyspnea: basis for once-daily and unequally divided twice-daily theophylline dosing schedules. Chronobiology International 1987;4(3):303–307.

81. Chan H-K, Young PM, Traini D, Coates M. Dry powder inhalers: challenges and goals for next generation therapies. Pharmaceutical Technology Europe 2007;19 (4):19–24.

82. Damms B, Bains W. The Cost of Delivering Drugs without Needles. Nature Biotechnology 1995;13(12):1438–1440.

83. Patton JS, Bukar J, Nagarajan S. Inhaled insulin. Advanced Drug Delivery Reviews 1999;35(2–3):235–247.

84. Uchenna Agu R, Ikechukwu Ugwoke M, Armand M, Kinget R, Verbeke N. The lung as a route for systemic delivery of therapeutic proteins and peptides. Respiratory Research 2001;2(4):198–209.

85. Ward EM, Woodhouse A, Mather LE, Farr SJ, Okikawa JK, Lloyd P, et al. Morphine pharmacokinetics after pulmonary administration from a novel aerosol delivery system. Clinical Pharmacology and Therapeutics 1997;62(6):596–609.

86. French DL, Edwards DA, Niven RW. The influence of formulation on emission, deaggregation and deposition of dry powders for inhalation. Journal of Aerosol Science 1996;27(5):769–783.

87. Ticehurst MD, Basford PA, Dallman CI, Lukas TM, Marshall PV, Nichols G, et al. Characterisation of the influence of micronisation on the crystallinity and physical stability of revatropate hydrobromide. International Journal of Pharmaceutics 2000;193(2):247–259.

88. Young PM, Cocconi D, Colombo P, Bettini R, Price R, Steele DF, et al. Characterization of a surface modified dry powder inhalation carrier prepared by "particle smoothing". Journal of Pharmacy and Pharmacology 2002;54 (10):1339–1344.

89. Martonen TB. Mathematical model for the selective deposition of inhaled pharmaceuticals. 1993; 82(12): 1191–1199.

90. Labiris NR, Dolovich MB. Pulmonary drug delivery. Part I: physiological factors affecting therapeutic effectiveness of aerosolized medications. British Journal of Clinical Pharmacology 2003;56(6):588–599.

91. Stone KC, Mercer RR, Gehr P, Stockstill B, Crapo JD. Allometric relationship of cell numbers and size in the mammalian lung. American Journal of Respiratory Cell and Molecular Biology 1992;6:235–243.

92. Hickey AJ, Thompson DC. Physiology of the airways. In: Hickey AJ, editor. Pharmaceutica Inhalation Aerosol Technology: Marcel Dekker, Inc.; 1992. pp. 1–27.

93. Pritchard JN. The influence of lung deposition on clinical response. Journal of Aerosol Medicine—Deposition, Clearance and Effects in the Lung 2001;14:S19–S26.

94. Shoyele SA. Controlling the release of proteins/peptides via the pulmonary route. In: Jain KK, editor. Drug Delivery Systems. Totowa, NJ: Humana Press; 2008. pp. 141–148.

95. Schanker LS, Mitchell EW, Brown RA Jr., Species comparison of drug absorption from the lung after aerosol inhalation or intratracheal injection. Drug Metabolism and Disposition 1986;14(1):79–88.

96. Kim HK, Chung HJ, Park TG. Biodegradable polymeric microspheres with "open/closed" pores for sustained release of human growth hormone. Journal of Controlled Release 2006;112(2):167–174.

97. Wang J, Chua KM, Wang C-H. Stabilization and encapsulation of human immunoglobulin G into biodegradable microspheres. Journal of Colloid and Interface Science 2004;271(1):92–101.

98. Yamamoto H, Hoshina W, Kurashima H, Takeuchi H, Kawashima Y, Yokoyama T, et al. Engineering of poly(DL-lactic-co-glycolic acid) nanocomposite particles for dry powder inhalation dosage forms of insulin with the spray-fluidized bed granulating system. Advanced Powder Technology 2007;18(2):215–228.

99. Wang J, Ben-Jebria A, Edwards DA. Inhalation of estradiol for sustained systemic delivery. Journal of Aerosol Medicine 1999;12(1):27–36.

100. Vanbever R, Mintzes JD, Wang J, Nice J, Chen D, Batycky R, et al. Formulation and physical characterization of large porous particles for inhalation. Pharmaceutical Research 1999;16(11):1735–1742.

101. Ben-Jebria A, Chen D, Eskew ML, Vanbever R, Langer R, Edwards DA. Large porous particles for sustained protection from carbachol-induced bronchoconstriction in guinea pigs. Pharmaceutical Research 1999;16(4):555–561.

102. Koushik K, Kompella U. Preparation of large porous deslorelin-PLGA microparticles with reduced residual solvent and cellular uptake using a supercritical carbon dioxide process. Pharmaceutical Research 2004;21(3):524–535.

103. Yang Y, Bajaj N, Xu P, Ohn K, Tsifansky MD, Yeo Y. Development of highly porous large PLGA microparticles for pulmonary drug delivery. Biomaterials 2009;30(10):1947–1953.

104. Salama R, Hoe S, Chan H-K, Traini D, Young PM. Preparation and characterisation of controlled release co-spray dried drug-polymer microparticles for inhalation 1: influence of polymer concentration on physical and in vitro characteristics. European Journal of Pharmaceutics and Biopharmaceutics 2008;69 (2):486–495.

105. Salama RO, Traini D, Chan H-K, Young PM. Preparation and characterisation of controlled release co-spray dried drug-polymer microparticles for inhalation 2: evaluation of in vitro release profiling methodologies for controlled release respiratory aerosols. European Journal of Pharmaceutics and Biopharmaceutics 2008;70(1):145–152.

106. Salama RO, Daniela T, Chan H-K, Sung A, Ammit AJ, Young PM. Preparation and evaluation of controlled release microparticles for respiratory protein therapy. Journal of Pharmaceutical Sciences 2009;98:2709–2717.

107. Liao YH, Brown MB, Jones SA, Nazir T, Martin GP. The effects of polyvinyl alcohol on the in vitro stability and delivery of spray-dried protein particles from surfactant-free HFA 134a-based pressurised metered dose inhalers. International Journal of Pharmaceutics 2005;304(1–2):29–39.

108. Dunne M, Corrigan OI, Ramtoola Z. Influence of particle size and dissolution conditions on the degradation properties of polylactide-co-glycolide particles. Biomaterials 2000;21(16):1659–1668.

109. Batycky RP, Hanes J, Langer R, Edwards DA. A theoretical model of erosion and macromolecular drug release from biodegrading microspheres. Journal of Pharmaceutical Sciences 1997;86(12):1464–1477.

110. Armstrong DJ, Elliott PN, Ford JL, Gadsdon D, McCarthy GP, Rostron C, et al. Poly-(D,L-lactic acid) microspheres incorporating histological dyes for intrapulmonary histopathological investigations. Journal of Pharmacy and Pharmacology 1996;48(3):258–262.

111. Müller RH, Maaen S, Weyhers H, Specht F, Lucks JS. Cytotoxicity of magnetite-loaded polylactide, polylactide/glycolide particles and solid lipid nanoparticles. International Journal of Pharmaceutics 1996;138(1):85–94.

112. Sivadas N, O'Rourke D, Tobin A, Buckley V, Ramtoola Z, Kelly JG, et al. A comparative study of a range of polymeric microspheres as potential carriers for the inhalation of proteins. International Journal of Pharmaceutics 2008;358 (1–2):159–167.

113. Asada M, Takahashi H, Okamoto H, Tanino H, Danjo K. Theophylline particle design using chitosan by the spray drying. International Journal of Pharmaceutics 2004;270(1–2):167–174.

114. Huang YC, Yeh MK, Cheng SN, Chiang CH. The characteristics of betamethasone-loaded chitosan microparticles by spray-drying method. Journal of Microencapsulation 2003;20(4):459–472.

115. Sakagami M, Kinoshita W, Sakon K, Sato J-I, Makino Y. Mucoadhesive beclomethasone microspheres for powder inhalation: their pharmacokinetics and pharmacodynamics evaluation. Journal of Controlled Release 2002;80(1–3):207–218.

116. Smith J, Wood E, Dornish M. Effect of chitosan on epithelial cell tight junctions. Pharmaceutical Research 2004;21(1):43–49.

117. Yamamoto H, Kuno Y, Sugimoto S, Takeuchi H, Kawashima Y. Surface-modified PLGA nanosphere with chitosan improved pulmonary delivery of calcitonin by mucoadhesion and opening of the intercellular tight junctions. Journal of Controlled Release 2005;102(2):373–381.

118. Learoyd TP, Burrows JL, French E, Seville PC. Modified release of beclometasone dipropionate from chitosan-based spray-dried respirable powders. Powder Technology 2008;187(3):231–238.

119. Learoyd TP, Burrows JL, French E, Seville PC. Chitosan-based spray-dried respirable powders for sustained delivery of terbutaline sulfate. European Journal of Pharmaceutics and Biopharmaceutics 2008;68(2):224–234.

120. Grenha A, Remuñán-López C, Carvalho ELS, Seijo B. Microspheres containing lipid/chitosan nanoparticles complexes for pulmonary delivery of therapeutic proteins. European Journal of Pharmaceutics and Biopharmaceutics 2008;69 (1):83–93.

121. Li FQ, Hu JH, Lu B, Yao H, Zhang WG. Ciprofloxacin-loaded bovine serum albumin microspheres: preparation and drug-release in vitro. Journal of Microencapsulation 2001;18(6):825–829.

122. Zeng XM, Martin GP, Marriott C. Preparation and in-vitro evaluation of tetrandrine-entrapped albumin microspheres as an inhaled drug-delivery system. European Journal of Pharmaceutical Sciences 1995;3(2):87–93.

123. Kwon MJ, Bae JH, Kim JJ, Na K, Lee ES. Long acting porous microparticle for pulmonary protein delivery. International Journal of Pharmaceutics 2007;333 (1–2):5–9.

124. Bosquillon C, Lombry C, Préat V, Vanbever R. Influence of formulation excipients and physical characteristics of inhalation dry powders on their aerosolization performance. Journal of Controlled Release 2001;70(3):329–339.

125. McConville JT, Patel N, Ditchburn N, Tobyn MJ, Staniforth JN, Woodcock P. Use of a novel modified TSI for the evaluation of controlled-release aerosol formulations. I. Drug Development and Industrial Pharmacy 2000;26(11):1191–1198.

126. Ho DH, Wang CY, Lin JR, Brown N, Newman RA, Krakoff IH. Polyethylene glycol-L-asparaginase and L-asparaginase studies in rabbits. Drug Metabolism and Disposition 1988;16(1):27–29.

127. Zalipsky S. Synthesis of an end-group functionalized polyethylene glycol-lipid conjugate for preparation of polymer-grafted liposomes. Bioconjugate Chemistry 1993;4(4):296–299.

128. Karathanasis E, Ayyagari AL, Bhavane R, Bellamkonda RV, Annapragada AV. Preparation of in vivo cleavable agglomerated liposomes suitable for modulated pulmonary drug delivery. Journal of Controlled Release 2005;103(1):159–175.

129. Garcia-Contreras L, Morçöl T, Bell SJD, Hickey AJ. Evaluation of novel particles as pulmonary delivery systems for insulin in rats. AAPS PharmSci 2003;5(2): Article 9.

130. Sharma A, Sharma US. Liposomes in drug delivery: progress and limitations. International Journal of Pharmaceutics 1997;154(2):123–140.

131. Allen TM. Liposomal drug formulations: rationale for development and what we can expect for the future. Drugs 1998;56(5):747–756.

132. Labiris NR, Dolovich MB. Pulmonary drug delivery. Part II: the role of inhalant delivery devices and drug formulations in therapeutic effectiveness of aerosolized medications. British Journal of Clinical Pharmacology 2003;56(6):600–612.

133. Thomas DA, Myers MA, Wichert B, Schreier H, Gonzalez-Rothi RJ. Acute effects of liposome aerosol inhalation on pulmonary function in healthy human volunteers. Chest 1991;99:1268–1270.

134. Alton E, Stern M, Farley R, Jaffe A, Chadwick SL, Phillips J, et al. Cationic lipid-mediated CFTR gene transfer to the lungs and nose of patients with cystic fibrosis: a double-blind placebo-controlled trial. Lancet 1999;353(9157):947–954.

135. Niven R, Pearlman R, Wedeking T, Mackeigan J, Noker P, Simpson-Herren L, et al. Biodistribution of radiolabeled lipid-DNA complexes and DNA in mice. Journal of Pharmaceutical Sciences 1998;87(11):1292–1299.

136. Smith PL. Peptide delivery via the pulmonary route: a valid approach for local and systemic delivery. Journal of Controlled Release 1997;46(1–2):99–106.

137. Wong JP, Yang HM, Blasetti KL, Schnell G, Conley J, Schofield LN. Liposome delivery of ciprofloxacin against intracellular *Francisella tularensis* infection. Journal of Controlled Release 2003;92(3):265–273.

138. Stark B, Andreae F, Mosgoeller W, Edetsberger M, Gaubitzer E, Koehler G, et al. Liposomal vasoactive intestinal peptide for lung application: protection from proteolytic degradation. European Journal of Pharmaceutics and Biopharmaceutics 2008;70(1):153–164.

139. Chono S, Tanino T, Seki T, Morimoto K. Efficient drug targeting to rat alveolar macrophages by pulmonary administration of ciprofloxacin incorporated into mannosylated liposomes for treatment of respiratory intracellular parasitic infections. Journal of Controlled Release 2008;127(1):50–58.

140. Bhavane R, Karathanasis E, Annapragada AV. Agglomerated vesicle technology: a new class of particles for controlled and modulated pulmonary drug delivery. Journal of Controlled Release 2003;93(1):15–28.

141. Cook RO, Pannu RK, Kellaway IW. Novel sustained release microspheres for pulmonary drug delivery. Journal of Controlled Release 2005;104(1):79–90.

142. Taylor KMG, Taylor G, Kellaway IW, Stevens J. The stability of liposomes to nebulisation. International Journal of Pharmaceutics 1990;58(1):57–61.

143. Darwis Y, Kellaway IW. Nebulisation of rehydrated freeze-dried beclomethasone dipropionate liposomes. International Journal of Pharmaceutics 2001;215 (1–2):113–121.

144. Kellaway IW, Farr SJ. Liposomes as drug delivery systems to the lung. Advanced Drug Delivery Reviews 1990;5(1–2):149–161.

145. Cortesi R, Esposito E, Luca G, Nastruzzi C. Production of liposomes as carriers for bioactive compounds. Biomaterials 2002;23(11):2283–2294.

146. Jaspart S, Bertholet P, Piel G, Dogné J-M, Delattre L, Evrard B. Solid lipid microparticles as a sustained release system for pulmonary drug delivery. European Journal of Pharmaceutics and Biopharmaceutics 2007;65(1):47–56.

147. Sanna V, Kirschvink N, Gustin P, Gavini E, Roland I, Delattre L, et al. Preparation and in vivo toxicity study of solid lipid microparticles as carrier for pulmonary administration. AAPS PharmSciTech 2004;5(2): Article 27.

148. Evora C, Soriano I, Rogers RA, Shakesheff KM, Hanes J, Langer R. Relating the phagocytosis of microparticles by alveolar macrophages to surface chemistry: the effect of 1,2-dipalmitoylphosphatidylcholine. Journal of Controlled Release 1998;51(2–3):143–152.

149. Yamamoto A, Yamada K, Muramatsu H, Nishinaka A, Okumura S, Okada N, et al. Control of pulmonary absorption of water-soluble compounds by various viscous vehicles. International Journal of Pharmaceutics 2004;282(1–2):141–149.

150. Yamada K, Kamada N, Odomi M, Okada N, Nabe T, Fujita T, et al. Carrageenans can regulate the pulmonary absorption of antiasthmatic drugs and their retention in the rat lung tissues without any membrane damage. International Journal of Pharmaceutics 2005;293(1–2):63–72.

8

Pharmaceutical development studies for inhalation products

Gaia Colombo[1], Chiara Parlati[2,3], and Paola Russo[4]

[1]*Department of Pharmaceutical Sciences, The University of Ferrara, Ferrara, Italy*
[2]*Department of Pharmacy, The University of Parma, Parma, Italy*
[3]*Novartis V&D, Technology Development, Siena, Italy*
[4]*Department of Pharmaceutical and Biomedical Sciences, The University of Salerno, Fisciano, Italy*

8.1 Introduction

Unlike the minimal trial-and-error approach, a systematic drug-product development process (also known as "quality by design," QbD) includes prior knowledge, the results of experimental designs, quality risk management, and vigilance during the product lifecycle. Such a systematic approach enhances final quality by including understanding and refinement of the formulation and manufacturing process in product development. Thus, the quality of an inhalation product has to be constructed during product and manufacturing-process development in order to meet patient needs and product performance.

The pharmaceutical development phase of a medicinal product's lifecycle (from idea to market) is when the product and the manufacturing process are designed, with the aim of consistently delivering the intended quality and performance [1,2]. An inhalation drug product is composed of at least one active pharmaceutical ingredient (API), a number of excipients—for formulation of the API and manufacture of the product—and a primary container (container closure system). The product quality and performance result from the integration and/or interaction of all elements after "assembly." However, it must be emphasized that the performance of an inhalation product is strongly influenced by the way in which it is manipulated, used, and stored once in the hands of patients and health care professionals.

Inhalation Drug Delivery: Techniques and Products, First Edition. Paolo Colombo, Daniela Traini, and Francesca Buttini.
© 2013 John Wiley & Sons, Ltd. Published 2013 by John Wiley & Sons, Ltd.

For every inhalation product, compendial texts such as pharmacopoeias and/or reference guidelines issued by regulatory agencies (European Medicine Agency (EMA), US Food and Drug Administration (FDA)) indicate the characteristics it must comply with to be used effectively and safely. Product characteristics, once identified and stated, enter as specifications in the main section of the product registration dossier (common technical document, CTD).

The characteristics of the product are identified and studied initially during the pharmaceutical development phase, following a rational design. These characteristics refer to drug substances, excipients, formulation, dosage form, container closure system, microbiological attributes, manufacturing process, and instructions for use—all critical to product quality. During pharmaceutical development, every critical formulation attribute and process parameter is studied for the extent to which its variation can impact on the quality of the drug product.

Before looking at the specific studies requested for inhalation products, it should be stated that for all drug products the correct pharmaceutical development strategy focuses on:

- components of the drug product (drug substance and excipients);

- the drug product (formulation development, overages, physicochemical and biological properties);

- manufacturing-process development;

- the container closure system;

- microbiological attributes;

- compatibility with reconstitution diluents (e.g. precipitation, stability) over the recommended in-use shelf life, at the recommended storage temperature.

The **drug substance** and **excipients** are studied for those aspects that influence the activity, safety, stability, bioavailability, and manufacturability of a drug product. Physicochemical and biological properties of the drug substance, such as solubility, water content, solid state, particle size, permeability, biological activity, and compatibility with excipients or other drug substances included in the same product, are evaluated. The type, concentration, and properties of excipients are also assessed, relative to their function. Tests are conducted to demonstrate their capacity to function as intended (e.g. as solvents, antioxidants, lubricants, penetration enhancers, release-controlling agents) throughout the drug product's shelf life.

As for the **drug product**, formulation development considers those attributes critical for the intended use and route of administration. These studies lead to the evolution of the formulation from the initial concept up to the final design, so long as the properties of drug substance, excipients, container closure system, dosing device, and manufacturing process have been optimized. In this phase, the use of an overage of drug substance (i.e. an amount of drug substance in the finished product higher than the labeled dose) is established and justified on safety and efficacy grounds. The overage can be requested to compensate for expected drug-substance losses during manufacture or product use (e.g. labeled numbers of doses in pressurized metered-dose inhalers (pMDIs) and device-metered dry powder inhalers (DPIs)).

Drug-product development studies allow for identification of the physico-chemical and biological properties relevant to product safety and performance, including formulation attributes and manufacturability. For example, studies have to include the development of a test for the respirable fraction (fine-particle mass) of an inhalation product.

The rationale for selecting the **container closure system** is based on the type of drug product and the suitability of the device for storage and transportation (compatibility with the content). In this regard, specific studies are conducted to justify the choice of materials for primary packaging, demonstrating the integrity of the container and closure. If the container closure includes a dosing device (e.g. DPI), studies are required to demonstrate that an accurate and reproducible dose of the product is delivered under testing conditions simulating the use of the product by patients.

Finally, **microbiological attributes** of the drug product must be considered during pharmaceutical development with respect to microbial limits for nonsterile products, the effectiveness and safety of preservative systems, sterility maintenance, and microbial challenge testing under conditions simulating patient use.

Development of the manufacturing process starts with consideration of the critical formulation attributes and the available manufacturing-process options, in order to select suitable process components and equipment. This allows any critical process parameter that should be monitored or controlled to be identified, in order to ensure that the product will have the desired quality. In general, development studies are conducted on more than one batch, thus accounting for batch variability.

In conclusion, the results obtained from pharmaceutical development studies combined with manufacturing controls create the so-called "product design space", that is the multidimensional combination/interaction of input variables (material attributes and process parameters) and output variables (product quality responses). Product variations within the design space are expected to remain under control.

The amount of data collected will be used to support the specifications proposed for the product, to be assessed routinely with every batch manufactured. Moreover, the development studies guarantee that those performance characteristics of the product that will not be routinely tested have been adequately investigated.

8.2 Pharmaceutical development studies for inhalation products

Inhalation drug products are "human medicinal products intended for delivery of the drug substance into the lungs, or to the nasal mucosa, with the purpose of evoking a local or systemic effect" [3,4]. These relatively complex means of dosage are typical combination products, comprising the formulation and a delivery device (Figure 8.1). The two combined elements "work together" for the efficacy of the therapy.

Inhalation therapy is successful so long as the active substance enters the airways at the expected dose and deposits on the respiratory mucosa. Upon deposition, a solid drug substance dissolves, is locally absorbed, and either exerts its effect onsite or becomes available for systemic distribution [5,6]. As the user activates the device, the drug formulation is transformed into a respirable aerosol of liquid droplets or solid particles, which can enter the airways by inhalation. The size distribution of the aerosol particles/droplets is the key characteristic determining the deposition site in the airways.

The delivery device coupled to the specific formulation is responsible for proper aerosol generation. In addition, in the case of multidose products, the delivery device also meters the amount of drug formulation that leaves the device at actuation. Hence, dose-metering and aerosol generation are two steps peculiar to inhalation-drug-product performance and are the focus of related development studies. Consequently, regulatory agencies demand with respect to the dose emitted (i.e. the amount of drug exiting the device and available to the patient) and the size

Figure 8.1   Inhalation pMDI and DPI products. Courtesy of Chiesi Farmaceutici, Parma, Italy

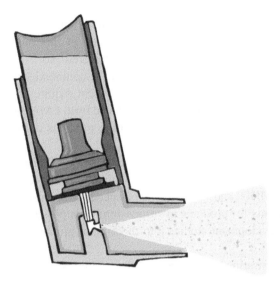

Figure 8.2   Sketch of pMDI function upon actuation

distribution of the aerosol mist (i.e. the aerosol respirability and the deposition site in the airways).

Inhalation drug products are categorized in four main types: pressurized metered-dose inhalers (pMDIs), DPIs, products for nebulization (single-dose or multidose), and metered-dose nebulizers (MDNs). DPIs can be further classified, as the dose can either be metered by the device at actuation or premetered during manufacture.

pMDIs consist of a drug substance formulated in suspension or solution in a liquefied propellant, stored in a metal canister sealed by a crimped metering valve, such that internally the formulation is under pressure (Figure 8.2). This internal pressure gives the energy needed to deliver and aerosolize the drug dose as the metering chamber is opened to the atmosphere; the propellant rapidly evaporates, producing the fine mist of solid particles or liquid droplets (depending on the formulation) that the patient inhales [7].

DPIs deliver a solid powder to the lungs [8,9]. The powder is either a micronized drug substance or, more commonly, a blend of micronized drug and larger carrier particles (such as lactose). In either case, the powder is delivered by means of a device in which the energy for delivery and aerosolization comes from the inspiratory flow of the patient's inhaling through the device.

Liquids for inhalation (solutions or suspensions) required "coupling" to an aerosolizing device (nebulizer) at the time of use. The nebulizer design and nebulization mechanism actually affect droplet size distribution, delivery rate, and total dose delivered. Since the formulation and device are often developed independently and various nebulizers are available on the market, the actual dose delivered to the patient remains uncertain and highly variable. However, in recent years technological improvements have been seen in this field, and the use of aerosolization in aqueous carrier is being reconsidered [10].

Given the variety of inhalation products and the complexity of their proper use and effectiveness, their pharmaceutical development requires many factors to be investigated and understood. Table 8.1, reproduced from EMA guidelines, provides a list of tests required to characterize inhalation products during pharmaceutical development [3].

As shown in the table, it is not necessary to perform all tests on every product. Some of them are applicable only to particular products (e.g. low-temperature performance for pMDIs) and some depend on specific label instructions for use (e.g. shaking requirements for products in suspension). It is possible to subdivide these pharmaceutical development studies into three main groups:

(1) Studies demonstrating product performance with respect to formulation, delivery device, and their combination.

(2) Studies demonstrating product performance in the hands of the patient (handling, use, storage).

(3) Additional studies demonstrating specific development requirements.

8.2.1   Studies demonstrating product performance with respect to formulation, delivery device, and their combination

Physical characterization

The physical properties of the drug substance and excipients may be relevant to the safety and efficacy of the product. These properties are solubility, particle size and morphology, density, surface rugosity, charge, and crystallinity [11–13]. For example, the dissolution of the drug substance in pMDI liquefied propellants makes a difference in product performance with respect to the particle size of the generated aerosol [14,15]. In fact, if the formulation is a suspension, the aerosol mist emitted upon valve actuation is composed of propellant droplets containing solid drug particles, whose size is expected to be the same established at manufacture. As the propellant evaporates, the solid drug particles already have a size suitable for inhalation and airway deposition. In contrast, if a drug solution is emitted, solid drug particles will be generated as a result of propellant evaporation, and their size (and shape) will depend on propellant evaporation rate, the presence of cosolvents, drug solubility, and so on. Clearly, this last process must be controlled to ensure that the size characteristics and degree of crystallinity of the newly formed drug particles meet the requirements for inhalation (respirability).

Table 8.1 Pharmaceutical development studies for inhalation products. White lines: studies demonstrating the product performance with respect to formulation, delivery device, and their combination; dark grey lines: studies demonstrating the product performance in the hands of the patient (handling, use, storage); light grey lines: additional studies for specific development requirements

	Pharmaceutical development study	Pressurized metered-dose inhalers (pMDIs)	Dry powder inhalers (DPIs)		Products for nebulization		Nonpressurized metered-dose inhalers
			Device-metered	Pre-metered	Single-dose	Multidose	
a	Physical characterization	Yes*	Yes	Yes	Yes*	Yes*	Yes*
b	Minimum fill justification	Yes	Yes	Yes	Yes	Yes	Yes
c	Extractables/leachables	Yes	No	No	Yes	Yes	Yes
d	Delivered-dose uniformity and fine-particle mass through container life	Yes	Yes	Yes	No	No	Yes
e	Delivered-dose uniformity and fine-particle mass over patient flow rate range	No	Yes	Yes	No	No	No
f	Fine-particle mass with spacer/holding chamber use	Yes	No	No	No	No	No
g	Single-dose fine-particle mass	Yes	Yes	Yes	No	No	Yes
h	Particle/droplet size distribution	Yes	Yes	Yes	Yes	Yes	Yes
i	Actuator/mouthpiece deposition	Yes	Yes	Yes	No	No	Yes
j	Drug delivery rate and total drug delivered	No	No	No	Yes	Yes	No
k	Shaking requirements	Yes*	No	No	Yes*	Yes*	Yes*
l, m	Initial and repriming requirements	Yes	No	No	No	No	Yes
n	Cleaning requirements	Yes	Yes	Yes	No	No	Yes
o	Low-temperature performance	Yes	No	No	No	No	No
p	Performance after temperature-cycling	Yes	No	No	No	No	Yes
q	Effect of environmental moisture	Yes	Yes	Yes	No	No	No
r	Robustness	Yes	Yes	Yes	No	No	Yes
s	Delivery-device development	Yes	Yes	Yes	Yes	Yes	Yes
t	Preservative effectiveness/efficacy	No	No	No	Yes**	Yes**	Yes**
u	Compatibility	No	No	No	Yes	Yes	No

*for suspensions.
**if preservative is present.

Particle size, shape, and density are combined in the definition of the aerodynamic diameter (d_{ae}). This equivalent spherical diameter describes the capability of aerosol particles to fly in the inhaled airstream, enter the airways, and deposit at a specific site [16,17]. Hence, size, shape, and density define the aerodynamic behavior of solid particles—either drug or excipient. In addition, for inhalation powders, particle size, shape, and density affect the flow and packing properties of the particle collection—that is, the powder. Dosage-form manufacture (drug/carrier blending, device-filling for multidose DPIs, capsule- or blister-filling for premetered single-dose devices) and product use (dose-metering, ejection from metering chamber/capsule, aerosolization) are strongly influenced by these fundamental and derived properties of the powder.

Drug particles for inhalation must be micronized with an aerodynamic diameter in the range 1–5 μm for respirability and lung deposition. However, the small size makes them poorly flowable, highly cohesive, and prone to aggregation. These drawbacks complicate the manufacturing phase while capsules or blisters are being filled with the correct drug dose, which by the way is often very low for inhaled APIs (microgram to milligram range). Moreover, a cohesive powder may be only partially aerosolized and emitted from the inhaler, thus causing a fraction of the dose to be unavailable to the patient. Again, if microparticles form aggregates of variable size, powder respirability and deposition will change, with unpredictable effects on drug bioavailability.

These are the reasons why flow properties should be improved, using several approaches [18–20] which change the physical appearance of the inhalation powders. For instance, many DPI formulations contain mixtures of micronized drug particles and larger lactose carrier particles, as shown in Figure 8.3. The carrier

20 μm

Figure 8.3 Scanning electron microscope microphotograph of an ordered mixture of lactose carrier particles and micronized drug particles. Courtesy Dr Francesca Buttini, Department of Pharmacy, University of Parma, Parma, Italy

improves powder-handling, dosing, and aerosolization [21]. Then, immediately upon inhalation, the blend is deaggregated due to the turbulence of the inhaled airstream: the micronized drug particles enter the airways, while carrier particles deposit in the mouth.

In alternative to ordered mixtures, highly respirable "engineered particles" can be designed by playing with their geometric size, shape, and density in order to define their aerodynamic size.

Whatever technology is chosen, the physical properties of drug particles, such as size, shape, density, and rugosity, will have to be assessed by means of suitable analytical tests (e.g. laser light scattering, electron microscopy, atomic force microscopy, etc.) during pharmaceutical development.

Finally, when a solid aerosol is inhaled, dissolution at the deposition site must occur for the active substance to become bioavailable. Thus, physical characterization must also consider solubility and crystallinity, as these may affect the rate of the dissolution process.

Minimum fill justification

Inhalation products exist in single-dose or multidose versions. Among multidose versions, the dose can be either premetered into a reservoir during manufacture or metered by the device at the time of use. Typical examples of multidose products are pMDIs and device-metered DPIs. These precisely meter a defined number of unit doses at each actuation. The number of available doses per container must be stated on the product label. During product use, the device indicates to the patient how many doses are still available by means of a dose counter [22]. The total amount of formulation (liquid or solid) to be filled into the container must be sufficient to provide the labeled number of actuations and is defined during drug-product development. This individual-container minimum fill guarantees that every dose of the total number labeled (particularly the final ones, delivered when the container is almost empty) meets the drug-product specification limits for delivered-dose uniformity and fine-particle mass (see the following sections).

Consequently, during product development, a study will be conducted to demonstrate the suitability of the individual-container minimum fill adopted. The study should prove that the filled amount is the minimum required to deliver all labeled doses. The extra fill added for the uniformity of the final doses must not induce the patient into prolonging use after delivering the last counted dose.

For premetered DPIs and products for nebulization, each unit dose is premetered by the manufacturer in a proper reservoir (capsule, blister, vial, ampoule), and the number of unitary containers in the box determines the labeled number of doses. In this case, the fill volume and/or weight inside the capsule or vial should be justified to comply with the specifications of delivered-dose uniformity and fine-particle mass.

Extractables/leachables

Determining the profiles of extractable and leachable compounds means evaluating any possible physicochemical interactions between the formulation and the container closure system [23]. The direct contact between content and container might lead to the presence in the drug product of "foreign" substances from the container. In the worst case, the drug product might be adulterated by a substance contained in the container (or a part of it) which renders it harmful. In this regard, regulatory expectations are higher for inhalation products than for other medicinal categories (e.g. topical, oral), as higher toxicological concerns relate to substance delivery to the lungs.

As a general definition, "**extractables**" are all compounds present in the container closure system that can be potentially extracted from it, independently of the drug product filled in. This means that extractables are specific to the materials (plastic, rubber) of which the container closure is made. On the other hand, "**leachables**" are a subset of extractables which can pass into the formulation under normal storage conditions during shelf life. Given a certain container closure, leachables will then be specific to each drug formulation.

Among the various types of inhalation product, the highest probability of an interaction exists in pMDIs, due to the presence of propellants and cosolvents that can extract into the formulation organic compounds from the valve or the canister coating. Therefore, assessing and quantifying the presence of leachables at trace levels (i.e. parts per million or billion) is a fundamental requirement during pMDI development. The issue is less relevant to DPIs, as an interaction between solids is rather unlikely to occur.

Packaging materials are selected at an early stage of product development, especially in the case of inhalation products where the container closure system contains and delivers the formulation. As for pharmaceutical excipients, for containers and closures compendial or noncompendial plastic and rubber materials are available to the drug-product manufacturer. Compendial materials are described in official pharmacopoeias. For this reason, their manufacturers (or the container manufacturers) are responsible for guarantying and certifying compliance with the specifications relevant for the intended use (including limits for extractables). Nevertheless, the justification and proof of any container-closure applicability is the responsibility of the drug-product applicant. It is required that the leachables profile for the compendial plastic and rubber components of the container closure selected for a certain formulation is determined.

For noncompendial materials that are in contact with the formulation (e.g. valves), the extractables profile will be determined by the drug-product applicant, and the conditions/results of the study will be provided in detail (e.g. solvents used, temperature, storage time). In addition, it is important to determine whether any of the extractables are also leachables found in the formulation at the end of the product shelf life or when the equilibrium point is reached.

A compound that appears as leachable should be identified and have its safety assessed based on suitable safety thresholds. Unfortunately, typical leachables "contaminate" the formulation at trace levels and the analytical procedures for their accurate identification and quantification can be complicated. In contrast, extractables can be present in the order of parts per thousand to per cent on the basis of material weight, and can thus be more easily detected, with positive results.

Delivered-dose uniformity and fine-particle mass through container life

The **delivered dose** is the amount of drug emitted in the form of an aerosol mist from the inhaler upon device actuation and available for inhalation.

The respirability of the aerosolized preparation is evaluated by the aerodynamic assessment of fine particles using a cascade impactor (see Chapter 6). The aerosol particles to be inhaled in the delivered dose must be formulated in such a way that a significant amount enters the lung. This part of the delivered dose is called the **fine-particle mass** and represents the amount of emitted aerosol particles suitable for lung deposition (i.e. with an aerodynamic size $<5\,\mu m$) [24].

Since the aerosol generated by an inhaler is a collection of particles, it is unlikely they will all have the same size; rather, they will show a more or less narrow size distribution, in the micrometer range. Basically, the fine-particle mass measures the respirable dose of the inhalation product. Development studies are aimed at improving the respirability of the inhalation product by increasing the fraction with a size suitable for deposition following inhalation. During pharmaceutical development, both delivered dose and fine-particle mass are studied in detail in order to guarantee their consistency according to drug-product application [25]. The goal is to minimize the difference between delivered dose and fine-particle mass.

The delivered-dose uniformity is experimentally measured by actuating the device a defined number of times and quantitatively collecting the emitted product into a specific sampling apparatus (see Chapter 6). The amount of drug in the sample is then quantified by a suitable analytical method (e.g. high-performance liquid chromatography, HPLC). The test is always conducted on the minimum delivered dose, which in some cases is emitted with more than one actuation.

Delivered dose and fine-particle mass are studied **throughout the container life**, from the first dose until the last labeled. The product performance must remain consistent regardless of whether the container is 100% full, half empty, or close to exhaustion. In general, a total of at least 10 doses has to be analyzed, combined from the beginning, middle, and end of the container's life.

For products which require priming before first use, the priming procedure must be studied and established to guarantee consistency of the first emitted dose. The product containers have to be handled and tested in the same way in which a patient would use them. Storage orientation, cleaning instructions, and dosing interval have to be considered. Moreover, a study should be conducted on the doses between the last labeled and the very last one at exhaustion. The rationale is to characterize the tail-off profile in case of incidental product use beyond the labeled number of doses (this test is not required if the product has a lock-out mechanism preventing such misuse).

Studies of delivered-dose uniformity and fine-particle mass are performed for all inhalation products designed to deliver a metered drug dose at each actuation (i.e. not for products for nebulization).

Delivered-dose uniformity and fine-particle mass over patient flow rate range

Delivered-dose uniformity and fine-particle mass must be studied in order to identify all variables that may affect them during product use. One main variable is the respiratory capacity of the patient, expressed in terms of inspiratory flow rate.

In the case of products for nebulization and pMDIs, the inspiratory airflow only captures the emitted aerosol, as the nebulizer and the propellant provide dose aerosolization and delivery. Hence, patient respiratory capacity is not an issue for delivered-dose uniformity and fine-particle mass, which only depend on the functioning mechanism of the device. Consequently, the product development is performed under a fixed airstream rate through the measuring apparatus.

In contrast, DPIs are devices in which the energy for dose aerosolization and delivery is given by the inspiratory act of the patient inhaling through the device. As the respiratory capacity and inhalation rate vary from patient to patient, this source of variability in product performance has to be addressed [26]. Hence, for DPIs, uniformity of delivered dose and fine-particle mass have to be studied over the range of flow rates achievable through the delivery device by the intended patient population. The device itself is also a variable to consider in defining the experimental conditions of this study, as its design determines the resistance offered to the act of inhalation through it. Consequently, for each device, a range of flow rates between 28 and 100 L/min must be identified (minimum, median, and maximum achievable) and justified based on clinical or published data. If the minimum flow rate is not sufficient to produce an acceptable delivered dose, information should be provided to health care professionals about the effect of inspiratory flow rate on product performance.

Fine-particle mass with spacer/holding chamber use

The spacer is a chamber mounted at the exit of the actuator of a delivery device that allows for the reduction of the aerosol emission speed in order to facilitate product use by the patient (Figure 8.4). The effect on fine-particle mass of the presence of a spacer/holding chamber only needs to be investigated for pMDIs requiring such an

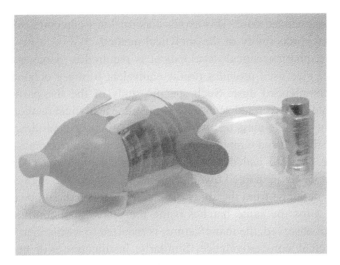

Figure 8.4 Spacers for children (left) and adults (right). Courtesy of Chiesi Farmaceutici, Parma, Italy

accessory. The spacer helps patients, particularly children and elderly people, in synchronizing device-firing with the inhalation act [27,28].

If a spacer is used, the emitted dose is aerosolized inside this chamber connecting the device to the patient's mouth. In this way, the coordination between device actuation and patient inhalation should be facilitated, while at the same time avoiding product loss in the environment. However, the presence of the spacer might change the fine-particle mass compared to that for the same dose emitted directly into the mouth. Fine-particle mass should thus be tested in the presence and absence of the spacer, in order to establish the instructions related to spacer use. Since a spacer has to be cleaned for successive uses (e.g. once a week), a study should be conducted both before and after cleaning it. Moreover, as patients behave differently when inhaling in the presence of a spacer (e.g. inspiration slightly delayed for 1–2 seconds, tidal breathing), the procedure should be modified to mimic this.

Single-dose fine-particle mass

In principle, the assessment of fine-particle mass should be conducted on the minimum recommended dose stated on the label. This would guarantee the consistency of product performance in terms of dose uniformity for every unitary smallest dose that the patient might receive. However, from a technical point of view, this is not always feasible—for example, when the minimum dose is very low (a few micrograms per shot) and the analytical method is not sufficiently sensitive. In that case, it is allowable for the routine test to use a sample size greater than the minimum recommended dose. This avoids abnormal test outcomes biased by poor accuracy and reproducibility of the analytical method.

In that case, pharmaceutical development is required to demonstrate that the routine sample size used provides results equivalent to those obtained using the minimum recommended dose. Such specific study of the fine-particle mass of one dose should be performed according to the same analytical method applied for the routine fine-particle-mass test without modifications, except those required to accommodate the reduced sample size. For example, in order to increase the amount of drug to be detected and overcome the analytical method's poor sensitivity, impactor stages may be pooled together prior to analysis.

The results obtained from the study of selected batches should be compared to those obtained for the same batches using the unmodified routine method. If differences are observed, the manufacturer is required to assess their significance and justify actual non-equivalence. Similarly, justification will be required if for any reason the test of fine-particle mass is not conducted on the actual single dose (e.g. for very low-dosed products). The test is not required for products for nebulization, as the concept of minimum recommended dose does not apply to them.

Particle/droplet size distribution

The size distribution of the generated and inhaled aerosol will determine the mechanism and level of drug deposition inside the airways, eventually affecting the success of the therapy [29].

Assessment of the size distribution of the liquid or solid aerosol generated by the inhalation device (combined with a specific formulation) applies to all product categories. According to the EMA guidelines on the quality of inhalation products [3], the determination of the particle size distribution of the dose fraction deposited on each impactor stage enables a complete aerodynamic characterization of the product used in the in vivo studies (pivotal, clinical, and/or comparative). Size-distribution data should then be provided for clinical batches as well as for batches representative of the commercial process, even though they may not be required routinely by product specification.

This type of in vitro development study is useful in establishing a quality reference for the product, keeping in mind however that impactors and test conditions greatly simplify (i.e. approximate) the whole inhalation process compared to what occurs in vivo [30,31].

For pMDIs and DPIs, particle size distribution is measured by means of multi-stage cascade impactors. These apparatuses measure the "aerodynamic size," which is a function of the physical dimensions of the particle ("geometric size"), but also of its density and shape.

Using a multistage impactor for size analysis, the results will show how much drug has deposited on each stage as a function of the particle size. It is recommended that the actual drug mass recovered from each stage and the cumulative mass below a given stage cut-off be determined, rather than the percentage of the delivered dose. The latter can hide variations in delivered dose.

A plot of cumulative percentage less than a stated cut-off diameter versus the cut-off diameter allows the determination of the **mass median aerodynamic diameter** (MMAD) and **geometric standard deviation** (GSD).

For nebulizers, droplet size distribution can be determined using cascade impactors or, alternatively, by laser diffraction, taking into account that changing the method of analysis may lead to different results for the same product [32].

Actuator/mouthpiece deposition

Actuators and mouthpieces are components of pMDIs, DPIs, and nonpressurized MDIs [33]. They are required to actuate the valve and release the dose from a pMDI, and to load every new dose before inhalation in a DPI. They are designed to fit a patient's mouth and are usually made of plastic.

Since the dose emitted from the canister in pMDIs or extracted from the capsule/blister/metering chamber in DPIs passes through the actuator/mouthpiece before entering the patient's mouth, it is important to determine the amount of drug depositing at this level. Such an amount is somehow lost—that is, not available to the patient—and must be taken into account in the definition of the ex-valve (or ex-delivery device) label claim, which corresponds to the delivered dose.

Delivery-device development

It is understood that for inhalation products, the formulation and delivery device, once combined, are equally important to the success of the administration [34]. For this reason, the development of the delivery device in combination with the designed formulation has to be carefully described. Hence, a slight change in formulation and/or device requires revaluation of the parameters critical to product performance. Whatever a change is implemented in the device design (e.g. component materials, actuator shape) and/or manufacturing process of the finished product, it should be discussed with respect to its impact on the product performance characteristics (e.g. delivered dose, fine-particle mass, etc.) [35].

For instance, for device-metered DPIs (Figure 8.5), developing the device means evaluating and preventing inadvertent metering of multiple doses and their subsequent inhalation by the patient. For all multiple-dose devices, a counter or fill indicator should be incorporated to give the patient a clear indication of when the total number of labeled doses has been delivered.

Figure 8.5 NEXT^{TM} DPI. Courtesy of Chiesi Farmaceutici, Parma, Italy

If the device is breath-actuated, the manufacturer should demonstrate that all target patient groups can trigger delivery. A detailed characterization of the trigger mechanism must be provided.

8.2.2 Studies demonstrating product performance in the hands of the patient (handling, use, storage)

As we have seen, many variables govern the success of therapy with an inhalation product. This is due to the product's complexity and the fact that the administration maneuver can be "unfriendly" for some patients. Clearly, appropriate performance depends on a product's robustness, but also on how its user (the patient) deals with it (Figure 8.6). Any perfect medicine can fail if not properly handled, used, or cared for by the patient [36]. Various aspects have to be considered in order to guarantee that the right drug dose reaches a patient's airways.

Cleaning requirements

It is recommended that inhalation devices are periodically cleaned during use, and device labels thus provide instructions on the method and frequency of cleaning. This is particularly important for nebulizers, as they are at risk of becoming vehicles of microbial infection [37,38]. However, it is also relevant to pMDIs and DPIs, since the deposition of product in the actuator or mouthpiece can affect the delivery of successive doses. Cleaning requirements and instructions are defined during pharmaceutical development by studying the effect of a clean or unclean actuator on delivered-dose uniformity, fine-particle mass, and droplet size distribution.

Figure 8.6 Product performance in human hands

These studies should be conducted under normal conditions of product use (e.g. priming, dosing intervals, and typical dosing regimen) in order to understand the relationship between cleaning and product performance.

Low-temperature performance and performance after temperature-cycling

Exposing the product to temperatures other than the recommended storage temperature may be an issue for pMDIs [39]. This is related to the presence in the formulation of the liquefied propellant, whose vapor tension depends on the temperature at which the product is stored or used. The propellant's vapor tension determines the internal pressure necessary for delivery and aerosolization of the formulation. A significant change in temperature may change the inner pressure, leading to variations in delivered and fine-particle doses. In addition, the rate at which the propellant evaporates upon aerosol emission can be different at low temperatures, changing the aerosol particle size.

Based on these considerations, the effect of low-temperature storage on the performance of a product is studied during development by storing containers in various orientations for at least 3 hours at a temperature below freezing (0 °C). The product is then tested to determine the number of actuations required before the dose meets the specification for delivered-dose uniformity and fine-particle mass. If poor product performance is seen at low temperatures, another study is required to define the method to be used and amount of time needed to warm up the container sufficiently for adequate performance (and this information is included in the instructions for use). If this study is not performed, instructions on cold temperatures should in any case be given to the health care professional and the customer.

Finally, temperature-cycling can affect the stability of the formulation: if the formulation is a solution, the dissolved drug may precipitate as a result of decreased solubility at low temperatures.

Similarly, product performance should be tested after submission (in various orientations) to temperature cycles between recommended storage conditions and a temperature below freezing (0 °C). In addition, for suspensions, temperature cycles should also vary between storage conditions and high temperatures (e.g. 40 °C), as high temperatures can affect stability (particle size of the dispersed fraction, dissolution). The storage time under each condition should be at least 24 hours.

Upon temperature-cycling, containers should be inspected visually for any evident defect and the product should be tested with respect to leaking, weight variations, delivered-dose uniformity, fine-particle mass, the presence of related substances (degradation products), and moisture content.

Effect of environmental moisture

The effect of environmental moisture on product performance should be investigated during development, given that the finished product may be stored and used in conditions of uncontrolled humidity. The effect of humidity is particularly relevant for DPIs, because an insufficiently protected dry powder can adsorb moisture from the atmosphere. Moisture adsorption leads to powder aggregation, affecting particle size, flow properties, dose-metering, and drug stability [40].

For premetered products in which the powder is loaded into capsules, it is also important to consider the brittleness of the capsules under various humidity conditions, because the capsule has to be opened by puncturing before delivery. A very dry atmosphere may lead to capsule partial-breaking, with powder loss inside the device before or during delivery.

A development study is also required for pMDIs, in which moisture can enter through the valve, depending on the formulation and the type of valve. Formulations containing ethanol as a cosolvent mixed with the liquefied propellant are much more prone to attracting humidity than those without ethanol. If moisture enters into the canister, it can lead to chemical or physical instability of the formulation (in the case of solutions and suspensions, respectively) [41].

Robustness

Even though an inhalation product is marketed with its own instructions for use, this does not guarantee that all patients will use it properly (Figure 8.6). Hence, the manufacturer is required to investigate product performance under conditions simulating use by patients [42,43]. Normal use should include activation of the device as frequently as is indicated in the instructions. In addition, the effects of carrying the device between uses, dropping it, or disassembling it should all be evaluated, as should the robustness of any lockout mechanism (to prevent misuse after delivery of the last labeled dose).

For DPIs in which the formulation is a powder mixture (e.g. micronized API + lactose), product robustness should also be checked with respect to the vibrational stability of the mixture, by submitting the product to vibrations similar to those that occur during normal transport and use. If vibrations influence mixture uniformity and, consequently, delivered dose and/or fine-particle mass, the significance of the observed variations should be discussed in terms of product safety and effectiveness.

8.2.3   Additional studies demonstrating specific development requirements

*Drug delivery rate and total drug delivered
(only for products for nebulization)*

This test is required to assess the complete delivery profile of the batches of products for nebulization used in in vivo studies (pivotal, clinical, or comparative). In order to know the actual drug dose available to the patient, the total amount of drug delivered and the delivery rate should be assessed using a validated method in the aerosol generated with the nebulizing apparatus and under the settings selected for the in vivo study.

Clearly, this study cannot be required for every batch on a routine basis, since solutions for nebulization are sold independently and the patient (or the health care professional) combines them with a certain nebulizing apparatus at home. The performance of the product is likely to be substantially dependent on the chosen nebulizer. Moreover, even if drug delivery rate and total drug delivered are known, the outcome of the therapy with nebulizers will remain greatly affected by the capability of the patient to accurately follow the prescription and respect the treatment time.

Shaking requirements (for pMDIs and products for nebulization)

Shaking before use is required for all products containing liquid formulations in suspension, in order to avoid sedimentation or flocculation from altering dose uniformity. A study is thus performed not only to demonstrate that the shaking instructions provided to the user are adequate, but also to check (by testing delivered-dose uniformity) whether excessive shaking leads to foaming and inaccurate dosing.

Initial and repriming requirements (for pressurized and nonpressurized MDIs)

The priming of a container consists in actuating the device and firing to waste a defined number of doses before the patient can take the first dose. Initial priming is necessary because in a brand-new product, initial actuations can be inaccurate with respect to drug content (for example, because the inner tube and metering chamber are empty). The recommended number of priming actuations is supported by a development study proving that such a number is sufficient for subsequent doses to meet the drug-product specifications for delivered-dose uniformity. The study should be conducted on containers stored in various orientations before initiation, indicating and justifying the length of storage prior to the study's conduction.

Another study should be conducted to support the length of time for which the product may be stored without being used after the initial priming. After this time, the product must require a repriming—a number of actuations before the device again complies with delivered-dose uniformity specifications. Again, the effect of container orientation should be considered, as well as the need to test products at different stages of container life (i.e. after the delivery of some, half, or almost all labeled doses).

Instructions concerning initial priming and repriming actuations should be provided to the health care professional and the customer.

Preservative effectiveness/efficacy (for products for nebulization and nonpressurized MDIs)

For products containing a preservative, a study (challenge test) is required to demonstrate the effectiveness/efficacy of the preservative system. The lowest preservative content that is effective in controlling microbial growth within acceptable limits should be demonstrated.

Preservatives are usually added to multidose aqueous formulations in containers that do not isolate their contents from the external environment and may undergo microbiological contamination. The study results should indicate correct product storage after first opening/use and during subsequent shelf life.

Compatibility (for products for nebulization)

If the product is to be diluted prior to administration, compatibility between drug formulation and diluent should be demonstrated for all possible diluents over the range of dilution proposed in the labeling. Compatibility should also be evaluated with respect to mixing two different nebulization solutions [44,45]. Assessed

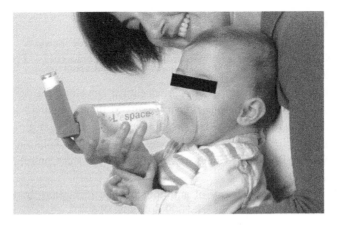

Figure 8.7 Inhalation therapy on a "special" patient. Courtesy of Air Liquide Medical Systems S.p.A., Bovezzo (BS), Italy

parameters should be precipitation, pH changes, droplet size distribution, drug delivery rate, and total drug delivered. The study should be conducted on aged samples and the sample should be followed throughout the entire storage time indicated for the diluted product.

8.3 Conclusion

It is clear that the aim of pharmaceutical development studies is to fully understand the drug-product characteristics that determine the efficacy, safety, and quality of a product. The optimal performance of the product depends on these features being integrated with the active role played by the user (Figure 8.7). The responsibility of the patient in using the product will be proportional to the level of complexity of both the product and the therapy [46,47].

Based on the information collected during pharmaceutical development, the manufacturer will be able to define the specifications of the drug product—that is, the list of features specific to the finished product that are to be tested routinely for each batch prior to release. Compliance with specifications at the time of release, as well as at the end of product shelf life, is for the patient a guarantee of quality, designed into the product during its development and continuously maintained under the unique responsibility of the manufacturer.

Acknowledgements

The authors are grateful to Chiara Simoni for drawing Figures 8.2 and 8.6.

References

1. EMA. Note for Guidance on Pharmaceutical Development. EMEA/CHMP/167068/2004. 2006.
2. FDA. Guidance for Industry—Q8 Pharmaceutical Development. 2006.
3. EMA. Guideline on the Pharmaceutical Quality of Inhalation and Nasal Products. EMEA/CHMP/QWP/49313/2005 Corr. 2006.
4. FDA. Guidance for Industry—Nasal Spray and Inhalation Solution, Suspension, and Spray Drug Products—Chemistry, Manufacturing, and Controls Documentation. 2002.
5. Traini D, Young PM. Delivery of antibiotics to the respiratory tract: an update. Expert Opinion on Drug Delivery 2009;6(9):897–905.
6. Carvalho RC, Carvalho SR, McConville JT. Formulations for pulmonary administration of anticancer agents to treat lung malignancies. Journal of Aerosol Medicine and Pulmonary Drug Delivery 2011;24(2):61–80.
7. Bell J, Newman S. The rejuvenated pressurized metered dose inhaler. Expert Opinion on Drug Delivery 2007;4(3):215–234.
8. Son YJ, McConville JT. Advancements in dry powder delivery to the lung. Drug Development and Industrial Pharmacy 2008;34(9):948–959.
9. Friebel C, Steckel H. Single-use disposable dry powder inhalers for pulmonary drug delivery. Expert Opinion on Drug Delivery 2010;7(12):1359–1372.
10. Watts AB, McConville JT, Williams ROIII. Current therapies and technological advances in aqueous aerosol drug delivery. Drug Development and Industrial Pharmacy 2008;34(9):913–922.
11. Chan HK. What is the role of particle morphology in pharmaceutical powder aerosols? Expert Opinion on Drug Delivery 2008;5(8):909–914.
12. Yin SX, Grosso JA. Selecting and controlling API crystal form for pharmaceutical development—strategies and processes. Current Opinion on Drug Discovery and Development 2008;11(6):771–777.
13. Hickey AJ, Mansour HM, Telko MJ, Xu Z, Smyth HDC, Mulder T, et al. Physical characterization of component particles included in dry powder inhalers. I. Strategy review and static characteristics. Journal of Pharmaceutical Sciences 2007;96(5):1282–1301.
14. Dalby RN, Byron PR. Comparison of output particle size distributions from pressurized aerosols formulated as solutions or suspensions. Pharmaceutical Research 1988;5(1):36–39.
15. Steckel H, Wehle S. A novel formulation technique for metered dose inhaler (MDI) suspensions. International Journal of Pharmaceutics 2004;284(1–2):75–82.
16. Glover W, Chan HK, Eberl S, Daviskas E, Verschuer J. Effect of particle size of dry powder mannitol on the lung deposition in healthy volunteers. International Journal of Pharmaceutics 2008;349(1–2):314–322.
17. Telko MH, Hickey AJ. Dry powder inhaler formulation. Respiratory Care 2005;50(9):1209–1227.

18. Pingali KC, Saranteas K. Practical methods for improving flow properties of active pharmaceutical ingredients. Drug Development and Industrial Pharmacy 2009;35 (12):1460–1469.

19. Jones MD, Harris H, Hooton JC, Shur J, King GS, Mathoulin CA, et al. An investigation into the relationship between carrier-based dry powder inhalation performance and formulation cohesive–adhesive force balances. European Journal of Pharmaceutics and Biopharmaceutics 2008;69(2):496–507.

20. Seville PC, Li HY, Learoyd TP. Spray-dried powders for pulmonary drug delivery. Critical Reviews in Therapeutic Drug Carrier Systems 2007;24(4):307–360.

21. Young PM, Kwok P, Adi H, Chan HK, Traini D. Lactose composite carriers for respiratory delivery. Pharmaceutical Research 2009;26(4):802–810.

22. Wasserman RL, Sheth K. Real-world assessment of a metered-dose inhaler with integrated dose counter. Allergy and Asthma Proceedings 2006;27(6): 486–492.

23. Norwood DL, Paskiet D, Ruberto M, Feinberg T, Schroeder A, Poochikian G, et al. Best practices for extractables and leachables in orally inhaled and nasal drug products: an overview of the PQRI recommendation. Pharmaceutical Research 2008;25(4):727–739.

24. Taylor A, Gustafsson P. Do all dry powder inhalers show the same pharmaceutical performance? International Journal of Clinical Practice, Supplement 2005;149: 7–12.

25. Chambers F, Ludzik A. In vitro drug delivery performance of a new budesonide/formoterol pressurized metered-dose inhaler. Journal of Aerosol Medicine and Pulmonary Drug Delivery 2009;22(2):113–120.

26. Martin GP, Marriott C, Zeng XM. Influence of realistic inspiratory flow properties on fine particle fractions of dry powder aerosol formulations. Pharmaceuticals Research 2007;24(2):361–369.

27. Lavorini F, Fontana GA. Targeting drugs to the airways: the role of spacer devices. Expert Opinion on Drug Delivery 2009;6(1):91–102.

28. Smyth HD, Beck VP. The influence of formulation and spacer device on the in vitro performance of solution chlorofluorocarbon-free propellant-driven metered dose inhalers. AAPS PharmSciTech 2004;5(1):E7.

29. Stein SW, Myrdal PB. A theoretical and experimental analysis of formulation and device parameters affecting solution MDI size distribution. Journal of Pharmaceutical Sciences 2004;93(8):2158–2175.

30. Mitchell J, Newman S, Chan HK. In vitro and in vivo aspects of cascade impactor tests and inhaler performance: a review. AAPS PharmSciTech 2007; 8(4):E110.

31. Newman SP, Chan HK. In vitro/in vivo comparison in pulmonary drug delivery. Journal of Aerosol Medicine and Pulmonary Drug Delivery 2008;21(1):77–84.

32. Waldrep JC, Berlisnki A, Dhand R. Comparative analysis of methods to measure aerosols generated by a vibrating mesh nebulizer. Journal of Aerosol Medicine 2007;20(3):310–319.

33. Kakade PP, Versteeg HK, Hargrave GK, Genova P, Williams R.C.III, Deaton D. Design optimization of a novel pMDI actuator for systemic drug delivery. Journal of Aerosol Medicine 2007;20(4):460–474.

34. Berger W. Aerosol devices and asthma therapy. Current Drug Delivery 2009;6 (1):38–49.

35. Mitchell JP, Nagel MW. Oral inhalation therapy: meeting the challenge of developing more patient-appropriate devices. Expert Review of Medical Devices 2009;6(2):147–155.

36. Lavorini F, Magnan A, Dubus JC, Voshaar T, Corbetta L, Broeders M, et al. Effect of incorrect use of dry powder inhalers on management of patients with asthma and COPD. Respiratory Medicine 2008;102(4):593–604.

37. de Vries TW, Rienstra SR, van der Vorm ER. Bacterial contamination of inhalation chambers: results of a pilot study. Journal of Aerosol Medicine 2004;17(4): 354–356.

38. Cohen HA, Kahan E, Cohen Z, Sarrell M, Beni S, Grosman Z, et al. Microbial colonization of nebulizers used by asthmatic children. Pediatrics International 2006;48(5):454–458.

39. Hoye WL, Mogalian EM, Myrdal PB. Effects of extreme temperatures on drug delivery of albuterol sulfate hydrofluoroalkane inhalation aerosols. American Journal of Health-System Pharmacy 2005;62(219):2271–2277.

40. Zeng XM, MacRitchie HB, Marriott C, Martin GP. Humidity-induced changes of the aerodynamic properties of dry powder aerosol formulations containing different carriers. International Journal of Pharmaceutics 2007;333(1–2):45–55.

41. Williams R.O.III, Hu C. Moisture uptake and its influence on pressurized metered-dose inhalers. Pharmaceutical Development and Technology 2000;5(2):153–162.

42. Nithyanandan P, Hoag SW, Dalby RN. The analysis and prediction of functional robustness of inhaler devices. Journal of Aerosol Medicine 2007;20(1):19–37.

43. Johnson GA, Gutti VR, Loyalka SK, O'Beirne KA, Cochran SK, Dale HM, et al. Albuterol metered dose inhaler performance under hyperbaric pressures. Undersea Hyperbaric Medicine 2009;36(1):55–63.

44. Akapo S, Gupta J, Martinez E, McCrea C, Ye L, Roach M. Compatibility and aerosol characteristics of formoterol fumarate mixed with other nebulizing solutions. Annals of Pharmacotherapy 2008;42(10):1416–1424.

45. Kamin W, Schwabe A, Kramer I. Inhalation solutions: which one are allowed to be mixed? Physico-chemical compatibility of drug solutions in nebulizers. Journal of Cystic Fibrosis 2006;5(4):205–213.

46. Rubin BK. Pediatric aerosol therapy: new devices and new drugs. Respiratory Care 2011;56(9):1411–1421; disc. 1421–1423.

47. Mitchell JP, Nagel MW. Oral inhalation therapy: meeting the challenge of developing more patient-appropriate devices. Expert Review of Medical Devices 2009;6(2):147–155.

9

Quality of inhalation products: specifications

Paolo Colombo[1], Francesca Buttini[1], and Wong Tin Wui[2]

[1]*Department of Pharmacy, The University of Parma, Parma, Italy*
[2]*Faculty of Pharmacy, Universiti Teknologi MARA, Puncak Alam, Selangor, Malaysia*

9.1 Introduction

The finished-product specifications define the quality that every product must exhibit in order to fulfill the requirements of safety and efficacy. These critical quality aspects are identified and studied during product development and the manufacturing process.

In the application dossier for market authorization, the quality necessary in order to go to market is set down in the Module 3, "Quality Data," part of the common technical document (CTD) (Figure 9.1). In this dossier, the qualitative and quantitative characteristics, test procedures, and acceptance limits with which the medicinal product must comply during its shelf life are detailed. The specifications are the acceptance limits applied to the tests performed on the finished product. They are a selection of the characteristics studied during pharmaceutical development and considered crucial to product quality. The characteristics of the batches of drug product used in pivotal clinical studies decide the specification limits [1].

The shelf life of the medicinal product is primarily established on the basis of the content of active constituents (efficacy), on the admissible level of breakdown products or impurities (safety), and on the consistency of pharmacotechnical properties (quality). The applicant for marketing authorization sets the specification limits at the time of batch release, such that the limits proposed at the end of the shelf life are guaranteed. These specifications at manufacture may be different from those at expiry.

Inhalation Drug Delivery: Techniques and Products, First Edition. Paolo Colombo, Daniela Traini, and Francesca Buttini.
© 2013 John Wiley & Sons, Ltd. Published 2013 by John Wiley & Sons, Ltd.

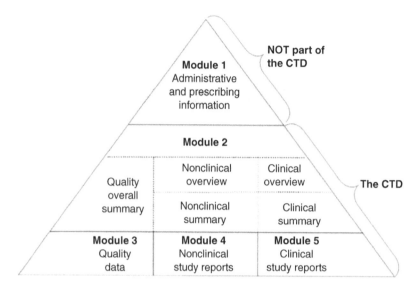

Figure 9.1 Diagrammatic representation of the organization of the common technical document (CTD)

An inhalation product is a solid or liquid drug formulation contained in a delivery device, intended for deposition in the respiratory tract. The inhalation product can deliver one or more drug, dissolved or dispersed in a suitable vehicle for administration, to the lungs as an aerosol. It is available as a single-dose or multidose, pressurized or nonpressurized, metered or device-metered product.

The therapeutic effectiveness of an inhalation product is dependent on the ability of the drug to be deposited at the intended site in the lungs for local and/or systemic effect. The safety and efficacy of an inhalation product is governed by quality attributes of the drug product, as well as the administration appropriateness and physiological and anatomical characteristics of the user's lungs. With reference to inhalation products involving metering, the dose delivered, the number of activations to provide the recommended dose, and the total number of activations per inhaler must be indicated on the label to ensure a safe, high-quality, and efficacious treatment is provided.

The medicine agencies dictate the specification tests and acceptance limits necessary for an inhalation product intended for market [2]. In Europe, this responsibility is allocated to the European Medicine Agency's (EMA) Committee for Medicinal Products for Human Use (CHMP), following consultation with the competent authorities of European Union (EU) member states and Health Canada. The specifications are communicated as scientific guidelines, incorporating the monograph information of the *European Pharmacopoeia* or equivalent. The guidelines harmonize the interpretation and validation of quality, safety, and efficacy aspects of inhalation products. They have been introduced with flexibility due to the range of inhalation products, which have considerable disparity in formulation and delivery-device characteristics.

9.2   Inhalation-product specifications

Pressurized metered-dose inhalers (pMDIs), device-metered or premetered dry powder inhalers (DPIs), single-dose or multidose nebulization products, and metered-dose nebulizers (MDNs), their manufacturing processes, container closure systems, drugs, and excipients, are all subjected to quality assessment.

Table 9.1 lists the specification tests for inhalation products in accordance with their dose and particulate, chemical, and microbiological properties. The listed tests are a combination of EU [2] and additional arbitrary selected US Food and Drug Administration (FDA) [3,4] requirements.

The quality attributes of an inhalation product must be maintained within approved limits from the date of batch release till the end of shelf life. As such, storage and handling conditions are based on the need to maintain the intended quality. The common tests required to qualify an inhalation product are description, dose/content, microbiological characteristics, chemical characteristics, and particulate characteristics (Table 9.1).

9.2.1   Description

A description of the formulation and the delivery device is essential to inhalation product recognition and as an indicator of integrity. The pharmacological effects of the drug and the excipient functions are also described. The description encompasses the composition and the appearance of the formulation and delivery device, which includes color, clarity, size, and shape. The actuator, immediate packaging material, and other parts which can influence drug delivery have to be described. In the case of products for nebulization, the immediate packaging is the container closure system, which for nebules (monodose solution) is the translucent low-density polyethylene ampoule.

9.2.2   Identification

The identity of the drug in an inhalation product is assessed with regard to a reference standard by use of techniques such as chromatography with ultraviolet or infrared spectroscopy detection. Complementary chromatographic procedures with a single integrated step, such as high-performance liquid or gas chromatography mass spectrometry, are used as well. The counter ion of a drug salt should also be identified. The identification methods should be specific in the case of a chiral drug.

Table 9.1 Specification tests for inhalation products. Tests in **grey** are those listed in [2]. + test requested; − test avoidable. © EMEA 2006 http://www.ema.europa.eu/docs/en_GB/document_library/Scientific_guideline/2009/09/WC500003568.pdf [Accessed 12th September 2012]

Specification tests		DPI		Nebulization products		MDN
	MDI	Device-metered	Premetered	Single-dose	Multidose	
Description	+	+	+	+	+	+
Dose/content						
Assay/drug content	+	+	+	+	+	+
Degradation products	+	+	+	+	+	+
Delivered-dose uniformity	+	+	+	−	−	+
Mean delivered dose	+	+	+	−	−	+
Content uniformity/ uniformity of dosage units	−	−	−	+	−	−
Minimum fill	+	+	+	+	+	+
Number of actuations per container	+	+	−	−	−	+
Leak rate	+	−	−	−	−	−
Weight loss	−	−	−	+	+	+
Microbiological characteristics						
Microbial limits	+	+	+	+*	+	+
Sterility	−	−	−	+**	−	−
Preservative content	−	−	−	+*	+*	+*
Chemical characteristics						
Moisture content	+	+	+	−	−	−
Leachables	+	−	−	+	+	+
pH	−	−	−	+	+	+
Osmolality	−	−	−	+	+	+
Particulate characteristics						
Fine-particle mass	+	+	+	+***	+***	+
Plume geometry/spray pattern	+	−	−	−	−	+
Extraneous particles	+	+	+	+	+	+

*Applicable to products with a preservative
**Applicable to sterile products
***Applicable to suspension products

9.2.3 Drug content

The amount of drug substance in the entire container is determined using a stability-indicating analytical method. The common assay limits of drug content for a medicinal product are ±5% of the drug in the label at the time of batch release.

The drug content can be analyzed using high-performance liquid chromatography or another assay procedure which allows the detection of degraded drug and changes in drug concentration in the formulation. A chiral assay is used to demonstrate that there is an insignificant racemization of chiral drug during manufacture and storage.

The drug content of a multidose inhalation product is expressed as the amount of drug per weight or volume unit. The drug content of a single-dose solution for nebulization is defined as the amount of drug per dosage unit [5].

9.2.4 Impurities and degradation products

Drug degradation products are impurities due to chemical changes during manufacture or storage, under the influence of light, temperature, or pH, or by reaction with the excipient and immediate container closure system.

The contents of a degradation product can be determined using the drug-content assay method, referring to the drug content in the product. Alternatively, they can be examined versus reference standards if the identity of the degradation product is known.

Both identified and unidentified impurities found in a threshold concentration of $\geq 0.1\%$ must be specified. Acceptance limits should be set for individual and total impurities. A stricter specification threshold may be adopted with an inhalation product administered at high daily doses, since the degradation products in a larger dose can potentially affect the therapeutic status of the medicine [2–4].

9.2.5 Preservative content

Those solutions intended for multiple administrations are added with preservatives in order to restrict microbial growth. This control is needed for nebulization products and MDNs which contain a preservative, a type of excipient, unless the preparation has an antimicrobial property. Preservatives are not required in dry inhalation products, due to the limited propensity for microbial growth in solid product. The preservative content of nebulization solutions and MDNs is analyzed using specific techniques. The acceptance criteria for product content should be based on the levels of preservative necessary to maintain the microbiological quality of the product.

9.2.6 Microbial limits

The presence of microorganisms in nonsterile products poses the threat of reducing or inactivating product activity, affecting the health of the user. Microbial examination of inhalation products is required in order to restrict the bio-burden of the dosage form. For inhalation products, the acceptance limits, expressed as colony forming unit (CFU)/g or CFU/mL, are 10^2 for total aerobic microbial count and 10^1 for total combined yeast/mold count. *Staphylococcus aureus*, *Pseudomonas aeruginosa*, and bile-tolerant gram-negative bacteria should be absent.

A microbial limit test is conducted through the incubation of a product sample in an appropriate agar or broth medium at specified temperatures and for specified durations, according to membrane-filtration, plate-count, surface-spread, or most-probable-number preparation protocols.

Casein soya bean digest agar/broth and Sabouraud-dextrose agar are the media employed in total aerobic microbial count and total combined yeast/mold count determination, respectively. Selective agar is used in the detection of specific microorganisms. In all microbiological examinations, a neutralizing agent may be added to remove the activity if antimicrobial agents are present [6,7].

9.2.7   Sterility

It is advocated that single-dose nebulization products be prepared in sterile conditions in order to comply with the sterility test (no evidence of microbial growth upon incubation of a product sample in suitable culture media for 14 days).

Fluid thioglycolate and Casein soya bean digest media are primarily used in culture of anaerobic bacteria, aerobic bacteria, and fungi. The product sample is introduced into culture medium following membrane filtration or through direct inoculation.

The probability of detecting microbes via sterility test increases with product sample number and the readiness of growth of any microorganisms present. The sterility test is a destructive assay and only the selected samples are tested. An appropriate sampling plan should therefore be adopted, considering batch size, volume of preparation per container, sterilization method, and so on. In the case of aseptic production, it is recommended that samples manufactured at the beginning and the end of the batch be selected, as well as samples made after significant manufacturing interventions [8].

9.2.8   Delivered-dose uniformity

The delivered-dose uniformity test is conducted according to a pharmacopoeia method, or a suitable validated alternative. Limits applied should be consistent with the pharmacopoeia, testing both intra- and interdevice variability. The mean delivered dose and delivered-dose uniformity of products for nebulization need not be assessed.

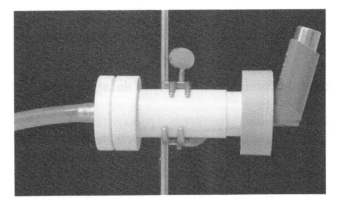

Figure 9.2 Set-up of a dose-collection apparatus with an MDI in place (from left to right: vacuum pump connection, filter, sample collector, adaptor, MDI)

It is however essential to characterize the uniformity of doses delivered from metered inhalation products, since the reproducibility of doses is affected by the actuation procedures.

The apparatus set-up for dose collection incorporates the sequential assembly of inhaler, mouthpiece adapter, sample collection body housing a filter, and vacuum pump connection (Figure 9.2, from right to left). The aspiration airflow rate created by a vacuum pump can be 28.3 L/min or higher.

A total of 10 doses discharged in the sample-collection body are assessed. The doses have to be collected over the life span of the inhaler—that is, at the beginning, middle, and end of the dose number stated on the label. The drug recovered from the dose actuated into the collection body is assayed. The delivered-dose-uniformity test must be conducted for each individual drug if the inhalation product contains more than one drug, and under breath actuation conditions if the inhaler is breath-operated.

The uniformity of delivered doses must comply with ±25% of the average dose for 9 out of 10 results. All should comply with ±35% of the average dose. If two or three values fall outside the range 75–125%, further tests on two more inhalers are required. In the latter, the uniformity of delivered doses must comply with ±25% of the average dose for 27 out of 30 results, and all should comply with ±35% of the average dose.

The delivered-dose uniformity of a solution formulation may be represented by uniformity of weight per actuation instead of uniformity of drug content of delivered doses. Justification of the reproducibility of drug content in doses delivered is requested in this case.

The uniformity of delivered doses among the containers of a batch must be assessed in addition to those from a single container. The sample size is 10 containers. The uniformity of dose per actuation between containers follows the same acceptance criteria applied to delivered-dose uniformity for a single container. A failed test should be repeated with an additional 20 containers [9].

9.2.9 Content uniformity/uniformity of dosage units

This test applies to solutions for nebulization packaged in a single-dose container. In the case of device-metered DPIs, MDIs, and MDNs, the dose-content uniformity is monitored using the previously described delivered-dose-uniformity assessment. Content uniformity should be investigated in samples removed from the containers as per the instructions provided to users and health care professionals. Acceptance limits should be justified, taking into consideration the pharmacopoeia requirements.

The content uniformity of the premetered DPI dose units should be controlled by a separate test. Typical acceptance criteria are prescribed in pharmacopoeia [9], such as a uniformity-of-dosage-units assay.

9.2.10 Mean delivered dose

In inhalation, the dose is a complex concept and attention has to be paid to the precise labeled dose. For pMDIs, DPIs, and MDNs, the content per actuation can be expressed either as metered dose—the quantity of drug contained in the device metering chamber—or as delivered dose—the quantity of drug available to the user, ex device. In the EU, all products containing new chemical entities or containing known drug substances used in inhalation products for the first time should be labeled with the delivered dose or an appropriate alternative (e.g. fine-particle mass). For existing products, current practice in each EU member state should be followed. In any case, it should be clearly stated if the label indication is expressed as metered dose (ex valve), delivered dose (ex actuator), or an appropriate alternative. Different products of the same drug labeled with the same metered or delivered dose might have a different therapeutic effect due to differences in the fine-particle mass.

The amount of drug in one actuation is determined as an average of doses delivered in a dose-uniformity test for a single container. In comparison to a drug-content assay, a wider limit of $\pm15\%$ of label claim is allowed to illustrate the drug content in one actuation. The mean delivered dose should be expressed as per actuation amount [9].

9.2.11 Number of actuations per container

In metered products with an actuator, the number of actuations for each container should be verified. MDI products can discharge up to several hundred metered doses of drug. Each actuation may contain anywhere from a few micrograms up to milligrams of drug, delivered in a volume typically between 25 and 100 mL. The number of actuations per container should be demonstrated to be not less than the labeled number.

Early fatigue of an actuator or of other related parts of the device, as well as a wrong setting in the locking mechanism on dose-counting, pose the risk of failure to deliver doses from the device during medication. A test can be conducted during testing of the uniformity of delivered dose. Doses should be discharged with intervals not less than 5 seconds between actuations, particularly in pMDIs. Excessively high frequency of actuation may lead to content freezing, and to valve or actuator blockage by frozen matter.

Dose-counting mechanisms are installed in metered-dose inhalers (MDIs) to indicate the remaining dose. In addition to the number of actuations per container, the functionality of the dose counter has to be examined [10].

9.2.12 Fine-particle mass

The aerodynamic particle size distribution of the delivered-dose aerosol depends on formulation, device, and patient maneuvers. Determination of the fine-particle mass is requested for all inhalation products and corresponds to the dose of drug assessed *in vitro* considered suitable for deposition in the lungs. Nebulization products should also be aerodynamically tested if they are formulated as suspensions.

The size of the particles in an inhalation aerosol can express the ability of the drug to reach the intended site of action. The optimum aerodynamic particle size distribution for most inhalation products is generally recognized as being between 1 and 5 μm. The fine-particle mass is then the amount of drug in an inhalation product that has an aerodynamic size considered capable of reaching the lung during inhalation (5 μm and smaller), on a per actuation basis. It is also called the respirable dose of an inhalation product and is often smaller than the delivered dose, due to formulation and administration limitations.

The fine-particle-mass test is conducted using a multistage impactor or impinger apparatus (see Figure 6.1) operating at 28.3 L/min or, in the case of DPIs, at a specific airflow rate dependent on the resistance of the device. The multistage cascade impactor fractionates and collects particles of one or more drug component by aerodynamic diameter through serial multistage impactions. A quantifiable drug mass, from a number of actuations not greater than 10, is discharged into impactor stages corresponding to different aerodynamic sizes of particles through the airflow. Aerodynamically large particles are deposited at the upper stages; aerodynamically small particles will follow the airstream and deposit at lower stages, when an adequate level of momentum is provided to them by discharge acceleration.

During the aerodynamic particle size distribution, the mass balance (total drug substance deposited on surfaces from the valve to the cascade impactor filter) is determined. The total mass of drug collected on all stages and accessories should be between 85 and 115% of label claim on a per actuation basis.

The fine-particle mass is computed as the drug amount pooled from stages corresponding to a particle size distribution less than 5 μm. However, a particle size distribution above 5 μm has to be controlled and analyzed as well, in case this fraction affects the therapeutic index of the inhaled product. Acceptance criteria for particle-size and size-distribution limits are proposed with the aim of characterizing the fine-particle mass of the dose.

The particle size and size distribution of discharged aerosol are described by the particle's mass median aerodynamic diameter (MMAD) and geometric standard deviation (GSD), calculated from a log-probability plot of the cumulative drug mass fraction against the cut-off diameter of various stages. In all cases, the proposed limits should be qualified by the fine-particle mass results for the batches used in *in vivo* studies (pivotal, clinical, and/or comparative) and should be reported on a per actuation or per dose basis. A total of five inhalers are used in fine-particle-mass tests, and only the initial discharge is quantified [11].

9.2.13   Spray pattern and plume geometry

A test of spray pattern and plume geometry is conducted during the pharmaceutical development phase of a spray inhalation product. Unless otherwise needed, the plume-geometry test can be excluded as routine for finished-product characterization. "Plume geometry" refers to the shape and size of the spray cloud, whereas "spray pattern" is the size and shape of the spray on a paper sheet.

The features of the spray pattern and plume geometry of a metered inhalation product indicate the performance of the valve and/or actuator. A deviation in spray pattern and plume geometry suggests that the uniformity of delivered doses is low and the ability of doses to reach the intended site in the lungs is poor, due to differences in the aerosol shot. Of interest is the difference in plume geometry from that using the same valve in CFC and HFA propellants. The Figure 9.3 shows that the plume of an HFA propellant, having lower pressure, is less elongated, indicating a slower plume motion [3].

The plume-geometry analysis of an aerosol proceeds by passing a laser beam through an individual spray plume, capturing over time high-speed pictures at 90° to the axis of plume. In a spray-pattern test, the actuation is performed on a plate (e.g. a chromatographic sheet), in order to collect the imprinting of the actuation. The actuation distance between the mouthpiece and the plate, the number of actuations, the position and orientation of the plate relative to the mouthpiece, and the visualization method are specified. The acceptance criteria for a spray

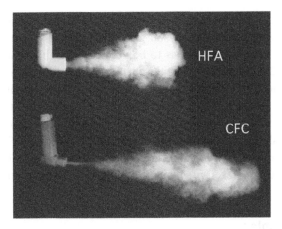

Figure 9.3 Plume geometry of HFA and CFC propellants. Courtesy of Dr Andrea Chiesi, Chiesi Farmaceutici, Parma, Italy.

pattern should include its shape (e.g. ellipsoid of uniform density) and size. The criteria can be set based on those of *in vivo*-tested inhalation products.

9.2.14 Leak rate

The leak rate of gas in pMDIs should be specifically determined, since drug-delivery performance is highly sensitive to variations in pressure in the canister. The leak rate indicates the pressure loss in the container during its shelf life. A leak-rate test should be performed in addition to the on-line leak test for occasional gross leaks, after the equilibration time before release.

Leakage of propellant concentrates the canister content of an inhalation product and can affect dose uniformity and particle size distribution. The leak-rate test is important because it provides information on pressure loss during storage and, therefore, on the stability of the product.

The leak rate is examined by gravimetric determination of 12 containers stored in an upright position at 25.0 °C and under constant humidity conditions, for two time intervals of not less than 3 days each. The leak rate is expressed as mg leakage per year of each container. The average leakage rate per year for the 12 containers should not be more than 3.5% of the net fill weight. None of the containers should leak more than 5.0% of the net fill weight per year [9].

9.2.15 Moisture content

A moisture-content test is critical for DPIs and pMDIs, as particles of these products are susceptible to aggregation by moisture. The aerosolization profile of DPIs is affected by moisture-induced particle aggregation. In pMDIs, particularly suspension formulations, polymorphism and crystal growth are linked to moisture content. A stable profile of moisture content has to be verified throughout the shelf life of an inhalation product, demonstrating that the changes in moisture content do not modify other product parameters. A weight gain with time suggests the possibility of moisture penetration into a product. When applicable, the moisture content can be assessed through product weight loss via infrared or oven heating. Alternatively, the moisture profile of an inhalation product can be determined by the Karl Fischer titrimetric method [12]. An upper limit for moisture content should be specified based on the inhalation product being used *in vivo* or else should be fixed at not more than 1%.

9.2.16   Leachables

Leachables are compounds which enter into a liquid or liquefied gas inhalation product from its container closure system under normal storage conditions. Leachable levels are dependent on the ability of fluids to interact with the container closure system and extract constitutive materials. DPIs, which have a negligible extraction capacity, are not tested.

The profile of the leachables from plastic and rubber container closure components—either directly or indirectly as volatile organic chemicals from the ink, paper, and adhesive of the label—is determined during pharmaceutical development of the inhalation product.

Routine monitoring of leachables is not necessary if there is no safety concern over the type and level of leachables detected. All tested inhalation products should adhere to the concentration limit set in the products used during *in vivo* trials [13,14].

9.2.17   pH

The characterization of the pH of an inhalation product is crucial as an abnormality in this attribute leads to adverse physiological responses. The acidic pH of the mucus lining the airways helps neutralize inhaled material and microorganisms. The pH of the nebulization solutions should not be lower than 3.0 or higher than 8.5. The mucus is able to buffer in some way the amount of solution deposited by inhalation. The pH of a liquid inhalation product must be characterized using a potentiometric instrument capable of reproducing pH values to within 0.02 units.

9.2.18   Osmolality

Like pH, the tonicity of an nebulization solution must be evaluated by measuring its state of osmolality. The osmolality of a nebulization solution can be determined from its freezing-point depression profile at the liquid state using an osmometer. It is expressed as milliosmole per kg of liquid. The blood has osmolality values ranging between 285 and 310 mOsmol/kg. The respiratory tract is nonetheless able to withstand deviations from the isoosmotic state of blood. In the *United States Pharmacopoeia*, a tobramycin inhalation solution has been formulated with osmolality between 135 and 200 mOsmol/kg. Osmolality limits for nebulization products lower than the dosage forms directly administered into the bloodstream are acceptable unless otherwise suggested by *in vivo* test performance of the inhalation product [15].

9.2.19 Extraneous particles

Extraneous particles in an inhalation product can derive from the manufacturing process, the drug, the excipient, and the container closure system. All inhalation products, but particularly nebulization solutions, should be subjected to tests for extraneous particulate contamination, remembering that levels of particulate matter in the drug product can increase with time, temperature, and stress.

A microscopy or laser-diffraction particle-size analyzer is employed according to the type of inhaler. There should be no more than 50 extraneous particles larger than 100 μm in an individual inhaler.

Microscopic examination of inhalation-product extraneous particulate matter in DPI formulations has the advantage of monitoring the morphology of drug and excipient particles, the extent of particulate aggregation, and any crystal growth. This information is important because polymorphism, crystallization, and/or aggregation of drug and excipient particles all affect delivery from the device to the lungs [2–4].

9.3 Additional quality aspects

9.3.1 Drug substances

"Drug substances" refers to new or existing actives described (or not) in pharmacopoeia. Sympathomimetic bronchodilators, glucocorticosteroids, antibiotics, DNA/RNA modulators, replacement surfactants, proteins, peptides, and anti-allergenic agents are all drugs which have been formulated into inhalation products for local and systemic uses.

The physical characteristics of a drug substance, such as solubility, particle size, shape, density, rugosity, charge, and crystallinity, may influence the homogeneity and reproducibility of the inhalation product. The careful physical characterization of a drug substance, relevant to its effect on the functionality of the product, is an important part of a development study.

The formulation of a drug as an inhalation product can transform its physical state from solid to liquid and vice versa. For solid drugs that do not undergo a dissolution process during manufacture, storage, or use, the characterization of drug particle size using a validated, multipoint particle sizing method is imperative. The median, upper, and/or lower particle size limits should be defined to ensure the particle size distribution of the drug is consistent with optimal size ranges for inhalation product formulation. Micronized drug substances are typically used in DPI or MDI suspensions. The particle size distribution and crystalline forms (e.g. shape, texture, surface) of the drug substance have to be carefully assessed. Other drug properties which may be monitored are color, visual and microscopic appearance, moisture, impurity, microbial limit, drug content, residual solvent, heavy metal content, and melting range. The acceptance criteria for a drug must be not less than those of injection products. In an inhalation product, the same drug can be supplied by alternative approved sources. Drugs from different sources have to be equivalent in their physicochemical properties, with special reference to aerodynamic particle size distribution [1].

9.3.2 Excipients

An excipient is a single chemical entity or a mixture of chemically related components. It can be obtained from a natural or a synthetic origin, and physically or chemically transformed into a technologically useful substance. In addition, it can be inert or can enhance the therapeutic effect of a drug. Like a drug, an excipient can be classified as novel, and can be described or not described in pharmacopoeia.

Propellants, cosolvents, diluents, antimicrobial preservatives, and solubilizing and stabilizing agents are typical excipients employed in inhalation formulation. Dehydrated alcohol as a cosolvent, lecithin as a surfactant, oleic acid as a lubricant, HFA 134 as a propellant, and lactose monohydrate as a carrier (Figure 9.4) are common excipients employed in inhalation products.

Excipients with a well-established record of use in inhalation products and tested in accordance with the standards of pharmacopoeia may be used without providing safety data. Toxicological assessments of an excipient are required when (a) it is used in inhalation above the previously employed amount, (b) it has been used in humans but never before by inhalation, and (c) it has not previously been used in humans. Physical, chemical, biological, and, if appropriate, immunological purity tests may be conducted, with limits for individual and total impurities.

In DPI products, multipoint particle-size analysis should be conducted on excipients, mixtures of excipients, and agglomerates of excipients and drugs. Again, the acceptance limit of such tests should be established from the results of batches used for *in vivo* studies, or from *in vitro* data using the multistage

Figure 9.4 SEM images of lactose particles: (a) smoothed lactose, (b) Lactochem by DFE Pharma, (c) Lactopress by DFE Pharma.

impactor. The physical properties of an excipient, such as polymorphism, crystallinity, impurity, moisture, surface roughness, and texture may be tested. Among such excipients with an effect on drug product performance, it is necessary to restrict the source to a single, validated supplier. Otherwise, the suitability of different suppliers must be demonstrated [16].

9.3.3 Container closure systems

The clinical efficacy of MDIs, DPIs, and MDNs is directly dependent on the design and performance characteristics of their container closure systems. The container closure system and the formulation collectively constitute the inhalation product.

In MDIs and MDNs, the container closure system consists of the container (canister), the valve, the actuator, and additional accessories (e.g. a spacer), as well as the protective packaging.

In DPIs, the container closure system consists of the overall device and all its primary and protective packaging. Since in DPIs the dose is extracted from the device by an inspiratory act, the device flow resistance is a peculiar parameter of these products (Figure 9.5).

A container closure system for an inhalation product should protect the product from microbial contamination, propellant/solvent loss, moisture/oxygen penetration, and light-induced drug degradation. In addition, in order to be compatible with formulation, it must be safe for patients, providing a reproducible performance in dose delivery. For a given MDI formulation, the design of the actuator, pump, valve, and other parts of the container closure system determines the delivery profile of the drug to the lungs. The chemical composition of the materials used in the construction of a container closure system must be tested, as this governs compatibility, safety, performance, and drug protection. Extractables and leachables in the container closure system must be studied in order to understand their

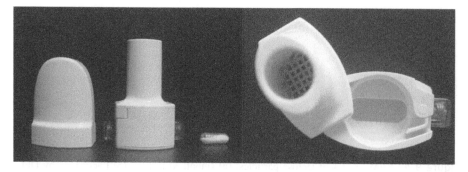

Figure 9.5   DPI device RS01/7, using a hydroxypropylmethylcellulose or gelatin capsule as a reservoir Courtesy of Plastiape IT.

physicochemical and clinical implications. Other crucial tests include moisture permeation, light transmission, microbial limit, sterility, and leak [13,14].

9.3.4   Inhalation-product stability

Inhalation products are subjected to stability tests in order to ensure their performance consistency with time across various temperature, humidity, and light conditions. The stability of an inhalation product upon manufacture, during storage, or during usage, is reflected by the physicochemical integrity of the drug and/or excipient. An important modification may change the delivery profile of the drug to the lungs, likely inducing adverse effects or negating the therapeutic effect.

The stability tests consist of long-term, intermediate, and accelerated studies in storage conditions characterized by different testing frequencies and durations. In the case of inhalation products, long-term, intermediate, and accelerated stability studies can be conducted at 25 °C/60% relative humidity, 30 °C/60% relative humidity, and 40 °C/75% relative humidity, respectively. Lower-temperature tests are employed in the testing of inhalation products stored in the refrigerator and freezer. Stability tests should be conducted on product attributes susceptible to changes during storage and likely to affect quality, safety, and efficacy. Product attributes such as physical, chemical, biological, and microbiological factors, as well as preservative content, weight loss, and aerosol functionality, are all examined.

A significant change in inhalation product stability is denoted by quantitative changes beyond the acceptance limits, the presence of nonacceptable degradation products, and the failure to meet appearance, physical-attribute, and functionality test criteria.

A specific concern with inhalation products (in particular MDIs) is the effect that storage orientation might have on product performance. During stability testing, the container should be placed in various orientations (0°: vertical upright; 45°: diagonal; 90°: horizontal; 180°: inverted upright) to verify the effect of orientation.

Quality attributes such as uniformity of delivered doses, fine-particle mass, and number of actuations per container have to be determined. Stability data should be presented in accordance with each container orientation.

In the case of an inhalation product using secondary packaging material for protection from light and/or humidity, such as a DPI, a stability test should be conducted on the unwrapped product in order to evaluate the duration limit of an accepted quality after the removal of packaging [17].

9.3.5 Package inserts and labels

Besides the general requirements, specific information addressed to users and health care professionals should be included in the "Summary of Product Characteristics" (a typical section of a registration dossier). The qualitative and quantitative composition, the posology and method of administration, and the conditions for storage are all mandatory in the package insert. In MDIs, DPIs, and MDNs, the content per actuation has to be clearly expressed as delivered dose or fine-particle mass: the drug amount making contact with the patient (Table 9.2).

This information is based on the outcome of specification tests, as well as experiments on product handling and performance. Inhaler priming and repriming in various orientations, rest time between actuations, handling temperature variation, cleaning method, and device robustness can all additionally be included. Storage conditions have to be stated, illustrating the effects of temperature, humidity, and light (Table 9.3).

The inhalation product performance indicators—namely delivered dose and particle/droplet size distribution—examined under varying airflow rates for patients of different age, gender, and disease severity, should be communicated. The goal is for users to adjust their administration practice. A guide on the wait time between actuation and respiratory act should also be given for inhalation products with a holding chamber or spacer.

With reference to the nebulization product, its output and droplet size distribution in association with specific nebulizer and operating conditions should be included as a package insert reference. Nevertheless, the consumer is free to select a nebulizer and operating conditions, which, however, can give rise to different dose and droplet characteristics.

9.3.6 Clinical/bioequivalence requirements

Finally, a few considerations on the therapeutic equivalence between inhalation products, in particular between an originator and the corresponding generic product, shall be presented.

Table 9.2   Specific information for the package insert of an inhalation product

Qualitative and quantitative composition

1. Name of product.
2. Name of drug.
3. Name of excipients.
4. Net fill weight.
5. Medication dose, delivered dose, metered dose, or fine-particle mass.
6. Number of deliveries from inhaler required to provide the minimum recommended dose, where applicable.
7. Number of deliveries per inhaler.
8. Delivered dose and droplet size distribution of the nebulization product using specific nebulizer and operating conditions.
9. Name of any added antimicrobial preservative. A warning statement relating to sensitization may be included.
10. Description of container closure system.

Note: a product which contains new chemical entity or known drug substance for the first time should be labeled with the delivered dose or an approved alternative such as fine-particle mass.

Method of administration

1. Shaking requirement.
2. Cold-temperature usage.
3. The number of priming sprays required before first-time use.
4. The number of priming sprays required for a unit which has not been used for a specified period.
5. The effect of flow rate on product performance.
6. Typical peak inspiratory flow through the device for patients with a wide range of pulmonary function.
7. Orientation of inhaler during inhalation.
8. Use of any specific spacer/holding chamber.
9. Cleaning requirement, including instruction for any spacer/holding chamber.
10. Step-by-step pictorial illustration of use.

Storage/usage precaution

1. The following statement is required for a pMDI:
   The canister contains a pressurized liquid. Do not expose to temperatures higher than 50°C. Do not pierce the canister.
   Do not use or store near heat.
   Do not throw container into fire or incinerator.

For an inhalation product:
   Avoid spraying into eyes.
   Keep out of reach of children.
   For oral inhalation only.
   Correct amount of medication in each spray cannot be ensured after the labeled number of sprays, even though the unit may not be completely empty.
   The number of sprays effected from each inhaler must be counted.

Table 9.2 *(Continued)*

The mouthpiece must not contain any foreign object before use or after removing the
protective mouthpiece cap, where applicable.

For a reusable inhaler device with a replacement cartridge or refill on the unit, not the device:
These units should be discarded when the labeled number of sprays have been dispensed.

For an inhalation product which requires protective packaging:
This product requires protective packaging to ensure its quality.
The product should not be used after a specified number of days following removal of the
protective packaging.
2. Preferred storage orientation.

Generic products must be essentially similar to innovator products. They must satisfy the criteria of having the same qualitative and quantitative drug composition and same pharmaceutical form, and of being bioequivalent to a particular innovator product. Therapeutic equivalence to the innovator of an inhalation generic product must be demonstrated by *in vivo* and *in vitro* studies. Therapeutic equivalence is defined as equivalent efficacy and safety (and quality).

Studies of clinical development are addressed to the investigation of the extent and pattern of pulmonary deposition of an inhaled drug substance. Regional quantification of the pulmonary deposition of two products can be carried out with imaging studies by measuring radioactivity in the different segments of the lung. In this case, three-dimensional scintigraphic methods are preferred. Classical

Table 9.3 Specific information for the label of an inhalation product

1. Name of product.
2. Name of drug.
3. Excipients.
4. Medication dose.
5. Delivered dose.
6. Net fill weight of the container.
7. Number of deliveries per inhaler.
8. Route of administration.
9. Allowable period of use after protective packaging is removed.
10. Shaking requirement for suspension product.
11. Recommended storage conditions, with warning.
12. Manufacturer's and/or distributor's name and address.
13. Lot number.
14. Expiry date.
15. "Rx only."

pharmacokinetics studies are useful for assessing pulmonary deposition, provided that the absorption of the active moiety from the gastrointestinal tract is excluded.

Pharmacodynamic studies are the alternative. Therapeutic equivalence demonstrated by means of appropriate clinical studies becomes mandatory when equivalence is not shown *in vitro* or through investigation of pulmonary deposition.

The bioequivalence status of a generic inhalation product is examined when delivery of this product to the lungs is accompanied by systemic drug absorption. For inhalation products which are intended to impact on local action, the delivered dose and particle size distribution of the generic product should be characterized against the innovator product. A complete particle size distribution of drug product from pMDIs, DPIs, and MDNs at each stage of a multistage impactor or impinger should be provided. The particle size distribution of a drug product may be assessed at various airflow rates and compared with an innovator product. Similarly, the droplet or particle size distribution of the nebulization product should be characterized. Comparison may be waived for generic nebulization solutions, which have the same qualitative and quantitative composition as an innovator product.

The draft guidelines on clinical documentation for orally inhaled products [18], including the requirements for the demonstration of therapeutic equivalence between two inhaled products for use in the treatment of asthma and chronic obstructive pulmonary disease, open the "therapeutic equivalence" with a reference product based only on the use of *in vitro* data. The product has to satisfy the following very strict criteria, imposing sameness, identity, and similarity:

- The product contains the same active substance (i.e. same salt, etc.).

- The pharmaceutical dosage form is identical.

- If the active substance is in a solid state (powder, suspension), any differences in crystalline structure and/or polymorphic form do not influence the dissolution behavior in the lungs.

- Any qualitative and/or quantitative differences in excipients do not influence the performance of the product (e.g. delivered dose uniformity, etc.), aerosol particle behavior (e.g. hygroscopic effect, plume dynamic and geometry), and/or the inhalation behavior of the patient (particle size distribution affecting feel).

- The inhaled volume needed to get a sufficient amount of drug is similar.

- The instructions for use of the inhalation device are the same.

- For breath-actuated inhalers: the inhalation device has the same resistance to airflow (within $\pm 15\%$).

- The delivered dose is the same (within $\pm 15\%$ of the labeled claim).

References

1. ICH Expert Working Group. ICH Harmonized Tripartite Guideline: the common technical document for the registration of pharmaceuticals for human use: Quality-M4Q (R1). Quality overall summary of Module 2. Module 3: quality. International Conference on Harmonization of Technical Requirements for Registration of Pharmaceuticals for Human Use; September 2002. http://www.ich.org/fileadmin/Public_Web_Site/ICH_Products/CTD/M4_R1_Quality/M4Q__R1_.pdf.

2. Committee for Medicinal Products for Human Use (CHMP). Guideline on the pharmaceutical quality of inhalation and nasal products. Committee for Medicinal Products for Human Use (CHMP), European Medicines Agency, EMEA/CHMP/QWP/49313/2005 Corr; October 2006. http://www.ema.europa.eu/docs/en_GB/document_library/Scientific_guideline/2009/09/WC500003568.pdf.

3. US Department of Health and Human Services, Food and Drug Administration. Guidance for industry: nasal spray and inhalation solution, suspension, and spray drug products—chemistry, manufacturing and controls documentation. US Department of Health and Human Services, Food and Drug Administration, CDER; July 2002. http://www.fda.gov/downloads/drugs/guidancecomplianceregulatoryinformation/guidances/ucm070575.pdf.

4. US Department of Health and Human Services, Food and Drug Administration. Guidance for industry: metered dose inhaler (MDI) and dry powder inhaler (DPI) drug products—chemistry, manufacturing and controls documentation. US Department of Health and Human Services, Food and Drug Administration, CDER; October 1998. http://www.fda.gov/downloads/drugs/guidancecomplianceregulatoryinformation/guidances/ucm070573.pdf.

5. Committee for Medicinal Products for Human Use (CHMP). Guideline on summary of requirements for active substances in the quality part of the dossier. Committee for Medicinal Products for Human Use (CHMP), European Medicines Agency, CHMP/QWP/297/97 Rev 1 Corr; February 2005. http://www.ema.europa.eu/docs/en_GB/document_library/Scientific_guideline/2009/09/WC500002813.pdf.

6. *European Pharmacopeia* 7.0. Vol. 1, Chapter 2.6.12. Microbiological examination of non-sterile products: microbial enumeration tests. Strasbourg, France: Council of Europe; 2011. pp. 163–167.

7. *European Pharmacopeia* 7.0. Vol. 1, Chapter 5.1.4. Microbiological quality of non-sterile pharmaceutical preparations and substances for pharmaceutical use. Strasbourg, France: Council of Europe; 2011. p. 507.

8. *European Pharmacopeia* 7.0. Vol. 1, Chapter 2.6.1. Sterility. Strasbourg, France: Council of Europe; 2011. pp. 153–156.

9. *European Pharmacopeia* 7.0. Vol. 1. Dosage forms. Preparations for inhalation. Strasbourg, France: Council of Europe; 2011. pp. 728–731.

10. US Department of Health and Human Services, Food and Drug Administration. Guidance for industry: integration of dose-counting mechanisms into MDI drug products. US Department of Health and Human Services, Food and Drug

Administration, CDER; March 2003. http://www.fda.gov/downloads/drugs/guidancecomplianceregulatoryinformation/guidances/ucm071731.pdf.

11. *European Pharmacopeia* 7.0. Vol. 1, Chapter 2.9.18. Preparations for inhalation: aerodynamic assessment of fine particles. Strasbourg, France: Council of Europe; 2011. pp. 274–285.

12. *European Pharmacopeia* 7.0. Vol. 1, Chapter 2.5.12. Water: semi-micro determination. Strasbourg, France: Council of Europe; 2011. pp. 140–141.

13. *European Pharmacopeia* 7.0. Vol. 1, Chapter 3.2.2. Plastic containers and closures for pharmaceutical use. Strasbourg, France: Council of Europe; 2011. p. 368.

14. US Department of Health and Human Services, Food and Drug Administration. Guidance for industry: container closure systems for packaging human drugs and biologics—chemistry, manufacturing and controls documentation. US Department of Health and Human Services, Food and Drug Administration, CDER, CBER; May 1999. http://www.fda.gov/downloads/drugs/guidancecomplianceregulator yinformation/guidances/ucm070551.pdf.

15. *European Pharmacopeia* 7.0. Vol. 1, Chapter 2.2.35. Osmolality. Strasbourg, France: Council of Europe; 2011. p. 57.

16. Committee for Medicinal Products for Human Use (CHMP). Guideline on excipients in the dossier for application for marketing authorization of a medicinal product. Committee for Medicinal Products for Human Use (CHMP), European Medicines Agency, EMEA/CHMP/QWP/396951/2006; January 2008. http:// www.ema.europa.eu/docs/en_GB/document_library/Scientific_guideline/2009/09/ WC500003382.pdf.

17. US Department of Health and Human Services, Food and Drug Administration. Guidance for industry Q1A(R2): stability testing of new drug substances and products. US Department of Health and Human Services, Food and Drug Administration, CDER, CBER; November 2003. http://www.fda.gov/downloads/drugs/ guidancecomplianceregulatoryinformation/guidances/ucm128204.pdf.

18. Committee for Medicinal Products for Human Use (CHMP) Guideline on the requirements for clinical documentation for orally inhaled products (OIP) including the requirements for demonstration of therapeutic equivalence between two inhaled products for use in the treatment of asthma and chronic obstructive pulmonary disease (COPD). CPMP/EWP/4151/00 Rev.1; October 18, 2007. http://www .ema.europa.eu/docs/en_GB/document_library/Scientific_guideline/2009/09/ WC500003504.pdf.

Index

Inhalation Drug Delivery: Techniques and Products, First Edition. Paolo Colombo, Daniela Traini, and
Francesca Buttini.
© 2013 John Wiley & Sons, Ltd. Published 2013 by John Wiley & Sons, Ltd.

Printed and bound by CPI Group (UK) Ltd, Croydon, CR0 4YY

27/10/2024

14580168-0004